Praise for

"Cox writes with grace an[...] amazing woman."   —*The Philadelphia Inquirer*

"This winsome memoir of a unique encounter with a baby gray whale off Seal Beach in California 17 years ago packs an emotional wallop."
                    —*The Washington Post Book World*

"A riveting adventure celebrating the mysterious bond between a champion swimmer and one wayward calf."
                    —*Elle*

"The story is thrilling enough, but it is Cox's poetic description of the ocean, the creatures she encounters and the life she sees all around her, that makes *Grayson* so captivating . . . As shimmering, colorful, and thought-provoking as it is refreshing: a reminder that our world has many hidden secrets, some of them just a few hundred yards offshore."
                    —*Long Beach Press-Telegram*

"Cox's flawless descriptions allow the reader to feel as if they're in the ocean with her."

—*Rocky Mountain News* (Denver)

"[An] endearing memoir and real-life fable."

—*Seattle Post-Intelligencer*

"*Grayson* would be compelling enough as a fable about a young woman and a lost whale. The fact that it's true makes the story wondrous, and unforgettable."

—Carl Hiassen

"Lynne Cox is a master of storytelling: her prose captures the vast movements and deep mysteries of the ocean and the creatures for whom it is home. Everyone who reads *Grayson* will be enchanted and profoundly moved. *Grayson* is a powerful voice for conservation."

—Jane Goodall

"A story of remarkable simplicity and charm. Through Lynne Cox's eyes we see an entire realm of creatures we have never known so intimately before. Truly for people of all ages. A parable and an experience, thanks not only to the author's great and generous spirit, but to her immense gift for describing nature."

—Anne Rice

"A beautiful true story of interspecies communication where the human and the whale mind connected."

—Temple Grandin,
coauthor of *Animals in Translation*

"It's safe to say that the woman with the world-renowned breast stroke has key strokes to match . . . [This is] a story of courage, stamina and remarkable compassion for another living creature . . . a story that only Lynne Cox could tell."

—*Union Leader* (Manchester, NH)

"If the heartwarming tale of a gentle sea creature separated from its mother doesn't envelop readers, Cox's description of the marine life will. Even those who aren't avid swimmers and dislike the ocean will enjoy a dip into Cox's watery world."   —*Cape Cod Times*

grayson

ALSO BY LYNNE COX

Swimming to Antarctica

grayson

LYNNE COX

MARINER

HarperCollins*Publishers*
Boston   New York

Mariner
An Imprint of HarperCollins Publishers, registered
in the United States of America and/or other jurisdictions.

www.marinerbooks.com

First U.S. edition published by Alfred A. Knopf, 2006

Library of Congress Cataloging-in-Publication Data
Cox, Lynne, 1957–
Grayson/Lynne Cox.—1st Harvest ed.
p. cm.
"A Harvest book."
Originally published: New York : A.A. Knopf, 2006.
1. Cox, Lynne, 1957– —Childhood and youth. 2. Swimmers—United
States—Biography. 3. Swimming—California—Santa Catalina Island—
Anecdotes. 4. Gray whale—California—Santa Catalina Island—Anecdotes.
5. Gray whale—Infancy—California—Santa Catalina Island—Anecdotes.
6. Human-animal relationships—California—Santa Catalina Island—
Anecdotes. 7. Wildlife rescue—California—Santa Catalina Island—
Anecdotes. 8. Santa Catalina Island (Calif.)—Description and travel. I. Title.
GV838.C69A3   2008
797.2009794'91—dc22   2007037895
ISBN 978-0-15-603467-8

Text set in Ehrhardt
Map illustration by Jeffery C. Mathison
Chapter opener illustrations by Gabriele Wilson

Printed in the United States of America

First Harvest edition 2008

23 24 25 26 27 LBC 23 22 21 20 19

To

David, Laura, and Ruth

ACKNOWLEDGMENTS

Over the last couple of years, I've discovered that a book is an enormous collaborative effort. Thank you to Vicky Wilson, my editor at Knopf, who believed in this story, and to Martha Kaplan, my agent, who encouraged me to write it and knew just where to place it. Thank you to the crew at Knopf, who transformed the manuscript into a beautiful book and helped it reach its audience; to Kenny Hawkins, my computer wizard; to Sherri Collins, who proofed the early versions of the manuscript; to the Rothwell family, who supplied great and constant love and energy; and to Linda Halker, who listened thoughtfully to the early drafts of the story. Also thanks to my

acknowledgments

family and friends who have supported me through the years to follow my dreams. Thanks to Dagmar for her inspiration and courage to overcome obstacles, great and small; to Cody, who stayed at my side for hours as I wrote the book and let me know when I needed walking breaks; and to Clara Kaplan, who helped me navigate through New York City and made sure I got to my editorial meetings. Thank you so much for all of your support.

grayson

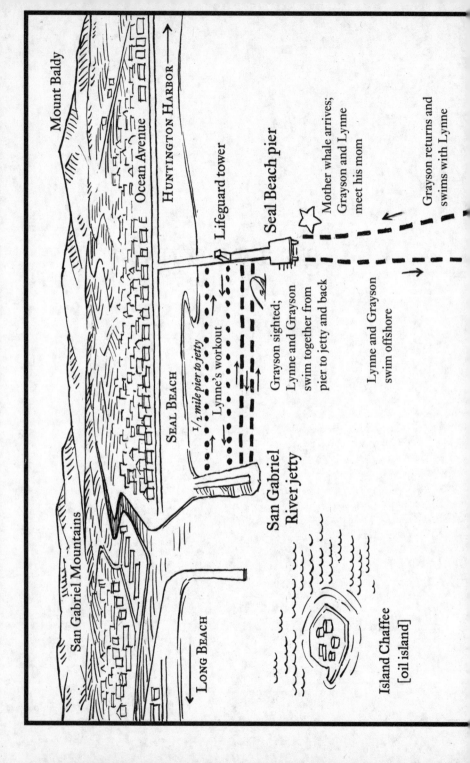

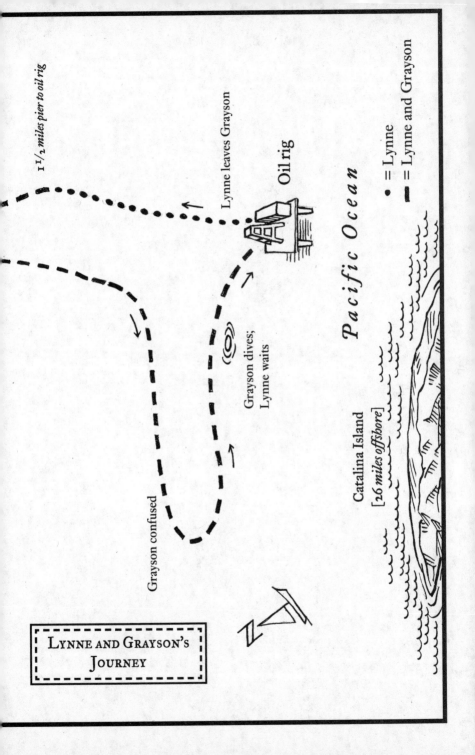

1 1/2 miles pier to oil rig

Lynne leaves Grayson

Oil rig

Pacific Ocean

Grayson dives;
Lynne waits

Grayson confused

Catalina Island
[26 miles offshore]

• = Lynne
▬ = Lynne and Grayson

LYNNE AND GRAYSON'S
JOURNEY

one

There's something frightening, and magical, about being on the ocean, moving between the heavens and the earth, knowing that you can encounter anything on your journey.

The stars had set. The sea and sky were inky black, so black I could not see my hands pulling water in front of my face, so black there was no separation between the sea and the sky. They melted together.

It was early March and I was seventeen years old, swimming two hundred yards offshore, outside the line of breaking waves off Seal Beach, California. The water was chilly, fifty-five degrees and as smooth as black ice. And I was swimming on pace, moving at about sixty strokes per minute, etching a small silvery groove across the wide black ocean.

Usually my morning workouts started at 6 a.m., but on this day, I wanted to finish early, get home, complete my homework, and spend the day with friends, so I had begun at 5 a.m.

There were vast and silent forces swirling around me: strong water currents created by distant winds and large waves, the gravitational pull of moon and sun, and the rapid spinning of the earth. These currents were wrapping around me like long braids of soft black licorice, and I was pulling strongly with my arms, trying to slice through them.

As I swam, all I heard were the waves, rising and tumbling onto shore, the smooth rhythm of my hands splashing into the water, the breaths that I drew into my mouth and lungs, and the long gurgling of silvery bubbles rolling slowly into the sea. I slid into my pace, and I felt the water below me shudder.

It wasn't a rogue wave or a current. It felt like something else.

It was moving closer. The water was shaking harder and buckling below me.

All at once I felt very small and very alone in the deep dark sea.

Then I heard a sound. I thought it was coming from the ocean's depths.

At first it seemed to be a whisper, then it grew louder, steadily, like someone trying to shout for help but unable to get the words out. I kept swimming and trying to figure out what was happening.

The sound changed. It became stranger, like the end of a scream.

In my mind, I quickly went through a list of the ocean sounds I knew and compared them with what I was hearing. There were no matches.

The hairs on my arms were standing straight out.

Whatever it was, was moving closer.

The ocean was charged with energy. It felt uncertain and expectant, like the air just before an enormous thunderstorm. The water was electric.

Maybe that was it; maybe the water was warning of an approaching squall. Maybe energy from distant winds and torrential rains was being transmitted through the water.

I checked the sky above and the distant horizon.

Both were dull and as black as ink and there wasn't a cloud in the sky.

I lifted my head to see the wave height. The shore break wasn't increasing and there weren't any wind waves. Not even dimples on the ocean's surface. There was no sign of a storm.

It didn't make sense. The energy in the water was intensifying. I felt like I was sitting on a tree branch beside a nest of angry, buzzing bumblebees.

All at once, the sea's surface erupted nearby. There was a rushing and plunking sound.

Like raindrops hitting the water. But nothing was falling from the sky. This was wrong.

Very wrong.

Out of the darkness, things were flapping into my face, flicking off my arms and head. It was like swimming through a sea of locusts, and with each impact my muscles tightened. I was tingling with fear, and all I wanted to do was to turn and sprint for shore.

But I told myself, Stay calm. You need to focus. You need to figure out what this is.

Taking a deep breath, I looked down into the deep black sea.

Thousands of baby anchovy were darting through the water like lit sparklers.

Blinded by panic, they were frantically tearing away from their schools and leaping out of the ocean like popcorn cooking on high heat. They were trying to evade something larger.

Light was exploding around me like hundreds of tiny blue flashbulbs constantly firing.

When I turned my head to breathe, something leaped into my mouth, wiggled across my tongue, and flapped between my teeth. It was larger than the water bug I once inhaled on a lake in Maine, larger than an anchovy.

Without thinking I spat it back into the sea. It had bright silver sides and was about six inches long. It was a grunion, a fish nearly twice as large as the baby anchovy. The grunion were chasing the anchovy, snatching them from the water and swallowing them whole.

More grunion were swimming in, bumping into my thighs, raking their pointy fins across my shoulders, but I smiled. The grunion had returned. Every year the grunion return to California in the spring and summer. They wait just offshore for the full moons or new moons when the tide is high, so they can swim ashore and lay their eggs. It always seems to be a miracle that they return every year and know exactly where and when to swim ashore.

A lone male grunion, a scout, swims ahead, and if the coast is clear, hundreds of female grunion follow him in, each with as many as eight male grunion swimming alongside. They choose a special wave, one that is on the receding tide so that it will carry them higher onto the beach, and the female's eggs will not be washed out to sea.

Once a female reaches the beach, she digs a hole in the sand with her tail, then wiggles back and forth, drilling herself down into the soft wet sand until she is buried all the way up to her lips. There she lays up to three thousand eggs, and one of the male grunion arches around her and releases his milt to fertilize the eggs. Then the adult grunion swim back to sea while the eggs incubate in the warm sand for ten days. Then the baby grunion hatch and ride the tide back out to sea to begin their lives in the ocean.

I loved to watch them come ashore and I loved to go

grunion hunting. It was a big event in Southern California. In summer, I would meet friends on the beach on moonlit nights and wait for the grunion. We'd spread our large bright-striped beach blankets on a berm, at the crest in the beach, beyond the reach of the incoming waves. We'd sit wrapped up in more warm woolly blankets, sometimes alone, or sometimes snuggled up with friends to stave off the cool, damp swirling ocean breezes. We'd talk, in muffled tones so no one would scare the fish away, about boyfriends and girlfriends, about summer plans and BBQs, about our lives and our families, our dreams and how we felt. We'd explore our lives, and sometimes touch hands under the blanket. We, too, were restless, awaiting our own high tide.

Someone in our group would whisper excitedly, "There he is!"

We'd jump to our feet, scanning the beach for a single fish. When we spotted one flopping on the sand, we'd watch and wait for what seemed like forever. Then a few minutes later, a wave would lift hundreds of grunion up. This wave would be so heavily laden with fish, it would rise more slowly than any other. As it curled, its dark glassy face would be altered by hundreds of grunion heads and tails protruding at all angles.

The wave would crash onshore and the grunion would spin and tumble across the sand, flipping, flopping, and pulling themselves beyond the water's edge. Their gills would beat in and out as they gasped for air. It seemed amazing to me that they could hold their breath for two or three minutes, and that they had to leave the sea and return to shore to continue the cycle of life. In utter fascination we'd watch this dance.

As soon as the grunion finished laying their eggs, they'd flip and flop back toward the water, and at that moment we'd charge across the sand, kicking mud on the backs of our legs and trying to scoop the grunion up with our bare hands.

They were always slippery, squirmy, and quick and harder to hold on to than a warm cube of butter. My friends and I might catch a few grunion, but none of us had the heart to take them home and cook them with a dusting of cornmeal and eat them as some people did. Somehow that would have spoiled the magic of all that we had witnessed. We were happy to catch them in our hands, feel the pulse of life racing through their bodies, and release them back into the warm salty waves.

As I swam I felt a strong connection with the agile schools of grunion and I thought I was lucky to be swimming with them—until I realized that they were attracting a small school of albacore tuna.

Usually the tuna lived and migrated twenty miles or more off the coast, but the abundance of food had lured them in. Albacore tuna are large fish. They weigh between twenty and forty pounds. They are shaped like giant oval beech leaves and have dark blue backs and gray-blue sides and bellies. They are very fast swimmers: they swim as if they are turbocharged.

At first I enjoyed feeling the way the water wavered and yawed as the tuna zipped to the right and left of me. But when they started leaping out of the water to catch the grunion, I grew concerned. I didn't want to be hit by a forty-pound tuna. I pulled to the right and then off to the left, but they were everywhere.

Then it happened. A big tuna weighing maybe thirty-five pounds rocketed out of the water. He smacked into my back and I jumped very high. Then another bounced off my shoulders. I started giggling. I had to roll on my side and catch my breath. It was raining tuna. What a weird, wild, and wonderful thing.

It occurred to me that these tuna would probably attract larger fish and the only larger fish I could think of were sharks. So I decided to move closer to shore, away from the feeding throngs. As I got nearer to land I started watching what was happening in the homes on the north side of the pier.

People were starting to get up. Second-floor windows that had been dark gray and vacant were becoming large glowing squares of gold, and as the people moved into their bathrooms and then downstairs more windows became gold squares. I imagined how warm it must be inside those homes. I let my mind enfold me in that golden warmth.

I was cold. The Pacific water temperature in March is in the mid-fifties; the surrounding water was constantly pulling heat from my body. It was like being wrapped up in a warm blanket on a snowy day and then having someone pull the blanket off. To overcome the heat loss, I had to swim at a rate fast enough to create heat, but still my skin always felt cold; it was as cold as the water. I could feel the cold working its way deep into my muscles.

An offshore breeze carried the warm sweet smells of smoky bacon and fried eggs, buttery pancakes, and the rich acidic aroma of brewing coffee across the water. I had been swimming for more than an hour and my stomach was grumbling loudly. All I had to do was reach the north jetty, turn around, and swim the last half mile back to the pier and then I'd be finished with my workout.

I was starting to relax, stretching out my arms, feeling my hands and arms pulling the thick water, feeling the rotation of my shoulders and core, and the light kick of my feet. My body was slipping through the water like silk sliding across ultrasmooth skin. My breaths were long and easy, and I felt good: I was back into my pace, moving with the flow of all creation. Everything was in sync, the currents flowing around me, the song of the ocean, the breeze—except everything below was strangely still.

All the fish had disappeared.

Lifting my head, I looked to my right and then to

my left. I couldn't see anything. I put my face back down, and stared into the water through clear goggles. It was like looking into a well at midnight. I couldn't see anything, but I knew something was there.

The water began shaking harder than before and I was being churned up and down as if I was swimming through a giant washing machine. The water shifted, and I was riding on the top of a massive bubble. It was moving directly up from below, putting out a high-energy vibration. I felt like there was a spaceship moving right below me. I had never felt anything this big in the water before.

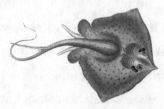

two

Turning sharply to the right, I tried to sprint for shallow water, but the creature diving below me was creating a huge hole in the water, and I had nothing to pull. There was no resistance. No support. I was tumbling into that huge hole, free-falling as if from the top of a cliff. I couldn't stop the fall, so I tried to bail out.

I spread my arms and legs out on the water to increase my drag, to increase my surface area and create more water resistance, but I was tumbling out of control, dropping deeper and deeper into the hole.

Every nerve was on high alert. My mind was trying to figure out what to do and my eyes were wide open, staring into the opaque black sea.

Pulling my hands rapidly and directly under my body, I tried to gain lift, enough to pull myself out of the hole back to the surface. But I was powerless.

Whatever was moving below was swimming so fast that it was dragging me along in its vortex.

Spinning my arms as fast as I could, I attempted to snap through the watery web. I kicked my legs as fast as I could move them. I rarely kicked. I used my legs mostly for balance—I was a lazy kicker—but adrenaline was shooting through my body, my heart was pounding, and my breath was quickening. My mind was racing.

What could be large enough to drag me along in its slipstream? Was it a California sea lion? They weighed up to two hundred and fifty pounds and grew up to six feet long. No, it felt a lot bigger than that, more powerful, faster, and it had displaced so much water. Was it a shark?

Seal Beach had always seemed like a safe place to swim. It was an area where I had worked out for years at all hours of the day and night. But things change. And I kept thinking, What is it? Is it a whale? What is lurking under me?

Holding my breath to slow my heart and the pounding in my throat so I could sense the slightest tremor in the water, I stretched out my right hand. The ocean currents were flowing under me, like a cool, constant breeze across a Nebraska cornfield. I tried to detect an irregularity. There were crosscurrents moving below, shifting under me like the winds across the northern plains. There was nothing erratic, nothing unusual.

Whatever it was had swum off—I hoped.

Then I lost control and focus. I took a couple of fast breaths and put my head down, sprinted toward shore. I couldn't keep doing this. I decided I had to swim just outside the wave break. It would be safer there. That way I could get out of the water quickly if I had to.

There was a splash—a huge splash. It was bigger than anything I'd ever experienced before. Large crescent waves bounced me up and down like the tail wave from a cigar boat moving at thirty miles per hour.

Instinctively, I turned and sprinted toward shore. I wanted to swim on the very edge of the surf line. I miscalculated.

I was slammed like a pancake between the griddle

and the spatula: A wave roared and crashed on my head, spun me horizontally side over side, hurtled me into the hard sand. Sand filled my swimsuit and lodged in my ears. Backwash tore me off the beach and dragged me backward into an approaching wave.

The lip of the wave caught me, pulled me deep into its mouth, chewed me and spun me so I couldn't tell up from down and then it flung me again into the hard-packed sand. It hurt.

Everything was a beat or two off. I couldn't get focused. My mind kept racing back to the question: What was swimming under me? I was scaring myself. Fighting with myself to stay in the water. I wanted to get out.

But I coached myself: You need to stick with it. Refocus. Get yourself back in gear. If you climb out now because you're scared you'll want to do the same thing when you're on a big swim. You need to do your workout so you build up your confidence and your strength, but you also need to be tuned in to what's going on around you. You need to be aware so that if something happens, you can respond immediately.

Psyching myself up, I stood up on the hard cold

sand just as another wave crashed on me, shooting saltwater up my nose. I stumbled to my feet, slid them along the bottom as fast as they would move and, when I reached waist-deep water, dove under the wave. I didn't like touching the bottom near the San Gabriel River jetty. It was dangerous.

There was an Edison power plant a couple of miles up the river. Water used to cool the turbines was dumped into the river. This warm water flowed down the river, swung around the jetty, and spread out along the shore. Here, the water was as much as ten degrees warmer. In winter you could see steam rising off the ocean. Stingrays loved the warm water. They congregated here, and it was a place where they had lovefests and multiplied to a point that the river jetty became a stingray city. This wasn't a thinly populated city like Cincinnati where there was plenty of room to spread out, room for spacious homes, porches, and big backyards. It was more like a stingray Manhattan, where the stingrays lived inches apart in the fine soft sand, and others lived right on top of one another in their own form of underwater apartments and condos.

Usually stingrays are docile. At the local aquariums

they lie in the touching tanks looking for humans. They push themselves up out of the water to have people pet them like puppies. Stingrays are flat fish that feel like a wet grape. They are light gray on top, white on the bottom, and they measure roughly two feet in diameter. Normally they bury themselves in the soft sand to hide from predators and look up at the world with two eyes on the top of their head. They have long tails that they use to propel themselves through the water.

At the end of those tails is a barb encased in a sheath. If a swimmer or wader inadvertently steps on a stingray, it will, in an effort to protect itself, whip its tail around and inject the barb and the sheath into the person's foot. The sheath has a protein substance on it that is very irritating, causing the foot to swell up two to three times larger than normal. Just as dangerous is the barb, which looks like an arrowhead with two spines pointing backward on either side. When the barb is injected into the foot, it locks into the skin; the only way to get it out is to have it surgically removed.

I was careful as I reentered the water, but I stepped on the edge of something. I felt it wiggle under my toes. It squirmed and I couldn't help myself: I screamed. I screamed really loudly. I never scream. And I jumped high out of the water. I wasn't thinking. Definitely wasn't thinking.

My feet were coming up off the bottom and before I knew it, I could feel them falling down. I wanted to stop them, but I wasn't thinking fast enough, wasn't able to get my mind to pull my feet back up. Even in the water, gravity was rapidly pulling them down.

My feet touched and I sank rapidly up to my calves in the fine mucky sand.

Things violently rammed into my legs. They were swarming and fluttering all around me like giant bats. I held my breath wondering if I would be stung.

Whatever I had stepped on had upset the entire colony of stingrays, and their neighbors too: guitar-fish, shovel-nosed sharks, and halibut that had been sleeping or hiding in the soft sand. The stingrays set off a chain reaction. The whole ocean floor was suddenly swimming with fish. On high alert, they were frantically trying to escape, bumping into one another and into me.

I wanted to jump, to pull my feet off the bottom as badly as if I'd been standing on red-hot coals. But I fought to keep my feet planted in the mushy silt, knowing that I wouldn't be stung if I didn't lift my feet and step down on anything. But when something swam right between my legs and I felt the upward push of its wings, I screamed louder than before.

It took all my focus to stand still and wait forever for a wave to break. It did, lifting me off the bottom: I kicked my feet up behind me and swam like mad through the surf.

When I made it beyond the waves, I noticed that my purple-and-white nylon suit was so filled with sand that it felt like I was carrying half of the beach with me.

This was one of the worst workouts I had ever had. But I told myself to deal with it, because when I did another channel swim, like attempting to break the world record for the Catalina Channel, I would need to be mentally prepared for anything. And this was preparing me for anything: That's why it was practice.

Bending over I pulled the bottom of my swimsuit to one side and let the large clumps of sand roll out of each leg hole. I started to swim again, but there was sand at the top of my suit too. And it was abrasive.

I was irritated that I had to stop again and get the sand out. If I didn't stop now it would soon be worse than running with a pebble in my shoe. The sand in my swimsuit combined with the motion of my arms would act like sandpaper rubbing my skin raw, but I didn't stop and remove it. Instead I pulled the front of my swimsuit open, kicked my legs rapidly, and let the saltwater wash the chunks of sand out of my suit.

Even though I knew it was important to stop and fix my suit at that moment, to take care of it so it wouldn't affect me in a bigger way later, I was annoyed with myself for stopping for so long. I needed to stay on my pace, to train the way I would do a channel swim.

Each day in the ocean was different. Each day I watched the wind move across the sea with giant brushstrokes and I'd anticipate that moment when the sun would slide above the horizon and I would watch the sunlight spread across the constantly changing surface of the sea. The intensity of the colors, of the reds, oranges, and yellows, would be magnified if it was a clear morning; on foggy mornings, the light would be soft pastel and fuzzy.

I took a deep breath, released the tension, and stopped to gaze into the sky. The earth was spinning closer to the sun.

The light was softening. The sky was changing from a shiny black to smoky granite gray and the sea was reflecting the change, a giant mirror to the heavens. It too was shifting from glossy black to wavery platinum. Taking in a very long deep breath, I relaxed a little more.

The promise of light made me feel a little more cozy and confident. At least now I would be able to see what was swimming under me—knowing was usually better than not knowing.

I lifted my head, checking for a fin. If the fin was sharp and angular, and if it was moving from side to side, it was a shark. If the dorsal fin was slightly curved and moving up and down, it was a dolphin.

Whatever had been swimming under me had seemed enormous. But my fear might have magnified the size of it.

Blue sharks, the non-human-eating species of sharks, rarely grew to more than ten feet long. White sharks were much larger, up to twenty-five or even thirty feet.

Local fishermen occasionally sighted great whites off Catalina Island, primarily on the west coast or what's known as the backside of the island—the place where the island is wide open to the Pacific Ocean and there's nothing but water until you reach Hawaii. I had never heard of a white shark sighting off Seal Beach, but there were no borders or barriers off Seal Beach to keep sharks out. They swam wherever they wanted to go. There was a large seal population offshore that rested on the large navigational buoys there, rolling with the tide and waves, their furry brown heads swaying from side to side. They slept on the rocks along the breakwater off Seal Beach and sometimes hauled out near the San Gabriel River jetty, north of the pier, just where I was heading. Seals were white sharks' favorite food. The Farallone

Islands off the shore of San Francisco were known as In and Out Seal Burgers for white sharks, which could jump out of the water and snatch seals off the rocks. It could have been a white shark cruising underneath me. It felt big enough.

three

The San Gabriel River jetty was only two hundred yards from me now. Once I reached it, I could turn around and swim back to the pier. I couldn't wait to finish this workout.

The sun was taking forever to rise. I was stuck in perpetual darkness. All I wanted was to get home, take a very long hot shower, and have something warm and delicious for breakfast.

Usually I loved swimming in the open ocean, but I was having a tough morning. And I couldn't shake the feeling that something really big was swimming nearby. I couldn't concentrate. I kept lifting my head to look for fins. I knew this was slowing me down;

lifting my head was causing my hips to drop and that created more drag. But I wanted to know what was swimming with me.

The urge to get out of the water became stronger. I wanted to get out, but I knew I had to make myself stay in and continue swimming: I reminded myself that I had to control my fear, otherwise I wouldn't be able to accomplish bigger goals later. I would have felt differently if I was certain there was an immediate danger.

Taking another deep breath, I swam with my head up and looked at the sky.

There were soft blue and yellow lights on top of the oil rigs on the islands off the shore of Long Beach, about three miles away. The oil islands were built to make the oil rigs more aesthetically appealing. The metal oil rigs were hidden behind walls, and water-falls were constructed to diminish the sounds of drilling. At night the structures and waterfalls were illuminated with blue, green, pink, and yellow lights and they looked magical. I used these lights as refer-ence points to help keep me swimming in a straight line. I gazed deeper into the blurry gray sky.

Turning my head to take a breath, I looked back under my arm at the eastern sky, hoping for sunlight.

Beyond the pier, the horizon was a thick black line where the sea and sky were pressed together like a giant eyelid, but high above a tiny yellow light flickered. There the horizon grew lighter, brighter, and softened. I rolled onto my back and slowly backstroked.

A wedge of red light parted the horizon.

The rose-colored sun moved slowly and majestically above the horizon. The giant eye was opening.

There was a discernible pause, a stunning peace, as if the sun and earth were silently shifting into sync. Then the sun climbed smoothly into the sky, casting a river of wavery rose light upon the water's surface. A long warm breath of wind ruffled the gray sea.

Seagulls stretched their wings, flapped them quickly, and cried loudly as they chased other gulls off the beach before lifting into flight and flying in a sweeping circle, squawking loudly as they headed toward a fishing boat. Tiny sanderlings raced in a big flock to the water's edge, hopping on one foot like they were jumping on pogo sticks. As they neared the water's edge, they dropped the other foot and scurried across the sand like wind-up toys, their tiny gray wings tucked against their backs. They chased the receding waves, whose long white lacy manes were glowing pink. The sanderlings poked holes in the sand with their short beaks, searching for sand crabs, while sandpipers trotted down from the high-tide line and stood on tall legs in knee-deep water, plunging their bills into softer sand in search of larger sand crabs. Seven pelicans flew overhead in single file, surfing the air currents created by the long rolling waves. Their wings were stretched out five feet wide, and they were underlit with rosy gold.

As the sun rose higher it turned tangerine and pushed the band of ruby red higher into the sky. The ocean resonated with color and warmth as I rolled back onto my stomach and swam across converging pools of red, orange, yellow, and gold.

Energy and warmth flowed across my back and shoulders. I was moving fast and free, feeling the power and lift of my arms and the strength deep within my body. My breathing was back to normal and finally this was fun again.

Pulling my hands directly under my body, I increased my lift in the water so that more of my back and legs was exposed to the sun's warmth. It would bake the chill out of my cold muscles.

On a breath, I swiveled my head around and looked over my right shoulder. Two strokes later, I breathed to my right side and watched the homes along Ocean Avenue slide behind me. My arm strokes were long and fluid. I slid on my stomach past the large pink Spanish-style house with the dark terra-cotta tiles, which marked the quarter-mile point. Just a quarter mile more to swim. I reached, pulled, pushed, reached pulled pushed, with each arm stroke past the streets perpendicular to Ocean Avenue, counting them off, Fifth, Sixth, Seventh, as I approached the Seal Beach pier, breathing every three strokes, listening to my bubbles roll into the water and to the rush of the waves like the breath of the sea. In and out we breathed together.

Glancing to my left, I watched the gray thirty-foot-tall wind-bent trunks of the palm trees that lined Eighth Street sway in the morning breeze; the clusters of dark green leaves on top of the trees were waving like a dozen hands, applauding the start of the day. And I wanted to applaud, too. It was almost over. I was almost there, almost finished with this long, cold, and difficult three-hour workout. I felt a sense of relief and a sense of accomplishment. I had been able to push myself, and stay focused, and complete the workout. And I was beginning to realize that I needed to not only prepare physically for something, but mentally as well.

The white clock tower with the bright orange tile roof atop City Hall read eight o'clock. With the morning's distraction, I was way off my pace: three minutes late. I was annoyed with myself. I told myself: Put your head down and sprint the last two hundred yards. Go. Go. Go. Pick up your pace. Pull stronger. Grab more water. Faster. Faster. Burn. Set the water on fire. Go. Go. Go. Reach for the stars. Go. Harder. You can do it. Ahh. Bring it home now. Go. Almost there. Yes, you've got it!

Reaching the pier, I rolled over on my back, sucking air. I swam backstroke slowly, trying to catch my breath. I was spent, cooling down, sore, tired, hungry, and eager to get home and finish my homework. I saw

Steve standing outside the bait shop. He was an old friend, a man in his sixties who ran the bait shop. He had worked there for as long as I could remember. He knew just about everybody and just about everybody knew and loved him.

Steve made a point of checking on me while I was swimming, especially in the darkness of early morning. He always tried to look nonchalant, as if he weren't watching out for me, but I could sense he was there. More than that: His silhouette beneath the soft white lights on the pier was easy to recognize. Steve always wore a short navy blue jacket that his broad shoulders filled out. He walked with an easy gait, even though his back was slightly bent with arthritis, and he was a little hard of hearing.

Whenever Steve saw me working out, he radioed the local commercial fishermen and the captain of the large boat that transported workers to the oil rigs that lay a mile and a half to nine miles offshore; Steve wanted to make sure they didn't run over me.

Usually I swam near "zero tower," the lifeguard station about halfway down the pier. That way I stayed well out of the way of boat traffic. My normal course was a half-mile stretch from the pier to the jetty and a half mile back again to the pier. Sometimes my workouts were three miles long and sometimes I swam as many as twelve miles. It depended on what I was training for. No matter how long the workout, I would always stop at the pier for at least ten seconds to catch my breath and check my lap time. And I always checked to see if Steve was outside.

I looked forward to seeing him. It made me feel like I had someone out there with me. It made me realize how much I appreciated him and how much I missed him when he had a day off.

Sometimes he'd be busy and just wave and sometimes I'd swim over to see him and talk. It slowed my workouts down a little, but I looked at it as a way to get a slightly longer rest, which gave me the motivation to swim faster on the next mile. We often joked. I loved to make him laugh. And I loved watching his silver gray head bob up and down, and seeing his mustache curl under his nose.

Steve knew more about the ocean than anyone I had ever met. He studied it every day, not out of a sense of duty, but out of curiosity and joy. He always wanted to learn something new and share whatever he discovered.

He spent a lot of his day swapping stories with local fishermen, with researchers and lifeguards. He saw the life, the seasonal changes, the natural signs and changing conditions within the ocean that no one else seemed to notice. He had a sixth sense about the sea.

Even in the blackness of early morning, Steve knew where I was in the ocean. He could spot me up to half a mile away. He saw through the darkness and across still or rough black water. He stood on the pier watching for the tiny neon blue sparks my hands made when they hit the water. There were zillions and zillions of light-emitting zooplankton and phytoplankton in the ocean.

When anything swam through the water—fish, seals, other marine mammals, or human beings—they left trails of bright shimmering light. The brightness of the light changed with the warmth of the water and the amount of plankton in the area. Sometimes the phosphorescence was so bright it was like looking deep into the Milky Way on a clear crisp winter's night, like swimming through a sky full of dazzling stars shooting across the black sky. And when there were fewer plankton in the water it was like swimming through the soft luminous light cast by Japanese lanterns. And sometimes, when the water was colder and the plankton was scarce, the trails were like the soft glow of candles in the distance.

Steve could tell how far I was from the pier by the size of the sparks my hands made when they hit the water; he knew my pace and when to expect my arrival.

When I looked at the pier, he wasn't where he normally stood. He was farther out. I knew something was wrong.

four

Steve was jumping up and down, rapidly waving his dark blue baseball cap and shouting. The morning breeze was tearing his words apart and carrying them away from me.

Cupping my ear, I gestured I couldn't hear him.

Quickly he pointed to something behind me.

Spinning halfway around I felt the water. Something was swimming under me. Was it a white shark?

Without hesitating, I sprinted for shore. Glancing over my right shoulder, I saw that Steve was vigorously shaking his head.

I stopped. I didn't want to. I was confused. What was he trying to tell me?

He cupped his hands around his mouth and shouted, "You can't swim to shore!"

"Why not?" I was baffled and wanted to get out so badly.

"That's a baby whale following you. He's been swimming with you for the last mile. If you swim into shore, he'll follow you. He'll run aground. The weight of his body on the beach will collapse his lungs and he will die."

"I don't see anything. Not even a fin," I said, searching the water.

Steve waved me over to him. When I was right below him, he explained, "Gray whales don't have dorsal fins. They have six to twelve knuckles along their back that they use for steering. I'm not surprised you didn't see the baby gray. They're hard to spot in the water. They're dark gray or black and they sort of blend in with the color of the ocean."

Early spring is when whales migrate, and the baby whale must have been swimming up from Mexico with his mother and somehow gotten separated from her. At that age they don't really know how to navigate effectively. Usually mother whales keep their babies very close to them and don't let them out of their sight. During their two- to three-month-long migrations the mothers let the young ones ride in their slipstream. By the time gray whales reach their summer feeding grounds in the Bering and Chukchi seas, the mother and baby will have swum between seven and eight thousand miles, swimming at a rate of two to six miles per hour. And in the autumn, they swim seven thousand miles back to Mexico, never sleeping and rarely eating, so that pregnant females can give birth in Baja where the water is warm and the lagoons are protected. These lagoons are great places for baby whales to learn how to swim. The male whales travel ahead of the mothers and their babies, who swim more slowly; eventually they all meet in the Bering Strait or the Chukchi Sea. When Steve told me whales travel up to fourteen thousand miles a year, more than halfway around the world, I was flabbergasted; I was tired after only a three-mile sprint.

He told me that gray whales are made to swim great distances so they can reach their summer feeding grounds in cold Arctic waters. The adults are filter feeders. They eat by moving along the ocean floor sucking in silt and fish, squeezing the sediment and water out through the baleen and capturing the tube-worms or shrimp.

But the baby grays can't feed like this. They are totally dependent on their mothers for the first eight months of their lives. They drink up to fifty gallons of milk throughout the day. This is their only food, so if they lose their mothers in the early months of their lives they will become dehydrated and then starve to death.

They need the milk for energy each day and also so they can put on the body fat that will keep them warm in Arctic waters. Their mother's milk is fifty-three percent fat, twice as rich as the richest ice cream.

Steve continued scanning the ocean for the baby whale. He said that gray whales were known as devil-fish. There are historical accounts of gray whales that when harpooned violently attacked the fishermen and their ships. Gray whales are also protective of their young. But if not threatened, they are huge, gentle beings.

Suddenly coming from the north was a long, loud *poof,* and then another *poof* and a third, louder *poof.*

It was the baby whale. He was about twenty-five yards from us. He was breathing: a white heart-shaped plume of mist was shooting, loudly, four feet above the water through the two holes on top of his head. The air smelled salty and oily, quite different from the breath of a minke whale, which smells like a cat's after a fish dinner.

Treading water, I watched the baby whale. I couldn't take my eyes off him.

He swam within ten yards of me. He was huge, maybe eighteen feet long, the length of a small sailboat. And he had to be at least three, maybe four feet wide. He was breathing deeply, spacing his breaths about fifteen seconds apart. The volume of air he exhaled made me even more aware of his magnificent size.

Lifting his huge fluke, the baby whale slid below the water's surface silently, like a cat stalking a bird.

Swimming under the water offered less resistance for him than swimming on the surface. Below the water, he swam more efficiently. His eyesight underwater was okay but it was limited by the clarity of the water. While gray whales do not echolocate like killer whales, dolphins, and porpoises, I wondered if they could still use their vocalizations as a crude kind of sonar to navigate. By vocalizing, some whales emit sound waves that bounce off whatever lies beneath the ocean. By listening to the sound waves that echoe back off the underwater objects, they can tell where they are underwater and what is around them. I also wondered if the baby whale could use his sonar to see in his mind's eye what he was hearing in the water—the way a musician reads music, so that he could hear and at the same time see what he was hearing, and then be able to distinguish the sounds more clearly.

The baby whale swam under me and I could feel waves of water peeling off his body and rolling under my legs and feet. Putting my face into the water, I looked down. He was about fifteen feet below me and if whales can smile I think he was smiling. He was moving his fluke (tail fin) up and down and slipping through the transparent gray water effortlessly with so much power and efficiency.

About two-thirds of the way down his back I noticed a small hump and then behind the hump near his fluke were six small knuckles. They really looked more like giant dimples. I wanted to touch them. And I wondered if the dimples on his body were like the dimples on a golf ball or like the ones on the wings of a plane that are made by severe hail storms. The dimples allow more wind to travel over the surface at a faster rate, which gives the ball and the plane more lift in the air. I wondered if the dimples on his back enabled water to travel over his body more quickly and if that gave him more lift than having a regular dorsal fin. I watched him swim.

He flew effortlessly through the water, rolling over underwater and making a slow giant spiral. Pointing his head up and dropping his fluke behind him, he kicked a few times, displacing so much water so quickly that he rocketed to the surface. Then he suddenly dove, rolling flipper over flipper like a crop duster doing wild and daring aerial maneuvers.

Despite his youth and size, the baby whale was in control of his flight through the water. He had learned to swim within an hour of his birth and it was obvious he was a natural. But he had refined his technique by practicing with his mother in the warm blue lagoons off Baja, Mexico. He knew just how far to kick his tail fin and when to hold his long soaring glide and how to alter his flipper position to turn precisely one way or the other.

The baby whale surfaced and immediately spouted. The spout rose four feet into the air. I couldn't help but smile. It was hard to believe that he was swimming so close to me. It was amazing to watch him. He seemed to be showing off or even trying to get me to play with him.

When he took a breath he floated easily on the surface. I watched the two holes on top of his head open and close. The lungs in his back must have been as large as two weather balloons. And that, plus his body fat, helped him rest on the water's surface and float with just the two holes in his head above the waterline so he could rest and breathe.

I wanted to get closer to see him, to see him better, to see what he was all about.

Turning his massive gray head, he looked directly at me with bright clear brown eyes the size of two enormous chestnuts. His face was mostly dark gray except for a couple of large white splotches near his chin, and he had vibrissae—small whiskers—on his face. The vibrissae were like a cat's whiskers, and he used them like a cat to sense what was around him.

He seemed to be gentle, but his size was intimidating. I'd never before been in the water with a being that was so large. I swam closer. I had heard that baby gray whales sometimes let people pet them and I really wanted to see what he felt like.

The baby whale rolled onto his side and floated. His body was breathtaking, perfectly streamlined. His mouth was large, stretching from one side of his head to the other and he held it slightly open, as if giving me a gentle smile. But he didn't open it far enough to enable me to see the baleen that he would eventually use to feed from the muddy ocean bottom. His head was large and his body was very elongated. In relation to the size of his body, his fluke and pectoral flippers were very short. He reminded me of a giant and gentle dachshund puppy. The fins on either side of his body were perfectly shaped, just like canoe paddles, with points at the tips that enabled him to capture the water down to the tips of his flippers.

He watched me swim breaststroke with my head above water. I swam slowly, keeping my eyes on him. I didn't want to scare him, or myself, but I wanted to check him out and see if he was healthy.

His skin looked smooth and clear and a lot like a gray wet suit, but it glistened in the sunlight. There weren't any fishing lines wrapped around his body; no debris was attached to him. He seemed to be physically all right.

I wondered if he was afraid. How scary it must be for him to be alone in the ocean, to be alone in such an enormous place. There were other fish and whales out there, but there was only one whale that was important to him. The one he depended on. The one he loved.

I wanted to reassure him so I swam closer.

The baby whale rolled over onto his stomach and the wave from his movement pushed me backward. He looked into my eyes as if he was trying to understand who I was and what I was doing there.

I was wondering the same thing about him and I had to ask in a soft voice so I wouldn't scare him, "What happened to you, little whale? Where is your mother? How did you get lost?"

If only I could speak his language. If only I could

find out what had happened. Most of all, I wanted to be able to tell him not to worry, that I would try to help. Two hearts in pursuit of the same thing were far stronger than one alone.

The baby whale knew this even though we couldn't speak. Something had brought us together; something much bigger than the two of us.

The whale dove and I pressed my face deep into the water so I could watch him. He was close, five feet from me. Holding as still as I could, I floated on my stomach. He didn't come any closer. He was so big. He seemed to sense that I was a little unsure of him. It surprised me that he didn't seem the least afraid of me.

He floated below the surface.

How do you do that? I wondered. I tried, but couldn't stay at his level. I popped back to the surface like a rubber duck.

He seemed to be listening to something, perhaps to some other whales somewhere nearby. Whales communicate at some frequencies that are too low for human beings to hear.

The baby whale inched closer.

"Don't worry, little one, we'll help you," I said underwater in a weird watery and garbled voice.

Lifting my head, I took a breath and looked up for Steve on the pier. He was leaning against the railing, shielding his light blue eyes with his hand, protecting them from the searing white sunlight.

He dropped his hand and said, "I don't see any sign of her."

"How do you think he got lost?"

"He is pretty young. He's between three and four months old. He may not have been listening to his mother."

Steve scanned the water again, swiveling his head from one side to the other.

At that moment, I realized how difficult it would be to find a whale in the ocean. Even something so big was actually so little in the vast sea.

The baby whale looked up at me through the water with his big brown doelike eyes. I felt something like a tingle, like the sound waves emitted by a wind instrument but without any music. I wondered if he was trying to communicate with me. Did he have senses that could tell him where he was? Could he use them to read what was in someone's heart?

"Do you think his mother's somewhere nearby?" I asked Steve.

"I do. I don't think she abandoned him. He looks healthy and seems to be swimming well. He's breathing without any difficulty. Something may have happened to his mother. She may have been injured."

I didn't say anything. I didn't want to imagine that. I wanted to believe that she was okay and we would be able to find her.

A thought is energy, and as it is transmitted, it is multiplied. Thoughts can either be positive, negative, or neutral, and they may travel all the way around the world as energy, affecting the way other people, and perhaps other beings, think. If I thought negatively, then I would put out negative energy, but if I thought positively, I would put out positive energy, expanding the possibilities of what could happen. It was very much like actors improvising: If they work together, stay in the moment, respond to one another in a positive way, they keep their skit going, moving forward, but as soon as someone puts forth something negative, the improvisation shuts down.

I needed to improvise, to stay in the moment, to remain positive, because I thought the baby whale would pick up on my energy. Maybe that's how he found me in the first place.

Steve understood this. He said, "The baby's mother has to be searching for him. She's probably calling him right now. A whale's vocalizations travel great distances under the water. She may be calling out his name, if whales have names. And I bet she's very worried."

"Do you think we'll be able to see her if she's in the area? How big do you think she'll be?" I asked.

"If she's nearby we'll see her. Male whales average thirty-five to forty-five feet long. Females are a few feet longer and they weigh between twenty and thirty-five tons.

"I think you should swim back to the jetty," Steve suggested. "That's probably where he lost her. Try swimming. He might follow you, like a puppy."

The baby whale was swimming near the pier pilings. Even though I was trying to keep the negative thoughts at bay, I didn't want to follow him under the pier. I didn't like swimming into the shadows. There were all sorts of sinister things under the pier, things that liked to reach out and grab you.

There was always fishing line, which often got tangled around the pilings, stretching across from one to another. The fishing line was invisible, so when I swam between the pilings, I could get tangled up in it. This freaked me out, especially when I felt an incoming wave. I knew that if I didn't get free of the lines the wave would smash me against the rough, dark brown wooden pilings, which were covered with white, razor-sharp barnacles and purple and black mussels. Both could shred the skin like a cheese grater.

Being under the pier made me feel anxious. The old fishing lines often had rusty hooks dangling right at face level. Worse than that was the colony of pier crabs that scurried sideways on the pilings over the barnacles and mussels with their arms stretched out, waving back and forth over their heads, ready to pinch anyone who got too close. Once a friend told me that a pier crab had pinched him. I didn't want to have a similar experience.

More than that, it took considerable skill to maneuver through the pier. There were five rows of pilings in some sections and four rows in others. In some cases it was better to go straight through the pier right between the pilings; in other cases, it was better to swim through on a diagonal. When a wave hit, it didn't matter which way you went. The key was to make it through without being rammed into a piling and really getting hurt.

When I saw the baby whale swimming toward the pier, I wanted to yell out to him, Don't swim there!

But he had no fear. Instead, he threaded his way through the pier, increasing his exposure time and possible danger.

But he made it through without a snag, and so I followed, riding a wave through the pier as he had done. And I laughed. It was so much fun. The more I tried, the more I could do, and if I listened and watched, I knew I could learn a lot from the whale.

Glancing back, I noticed a dark navy line of water paralleling the beach. The wave was building, increasing in height as the bottom of it hit the ocean floor. The wave hit the pier with so much force it shook.

The wave grew to five feet, caught the baby whale, suspended him in the air, and propelled him toward the beach like a flying log.

All I could see was the breaking back of the wave. And again I wanted to warn him: Watch out for that wave or it will beach you. You have to swim fast to get outside the break.

But the whale simply dropped his fluke, so he was vertical in the water, and used his tail like a giant brake, immediately stopping his forward momentum. He bailed out of the wave before it crashed and swam effortlessly toward me.

And as he swam, he was immersed in the water. He was one with it and his swimming motions came from the core of his body. His head moved down into the water, the top of it tracing a U. His body followed his head until he reached the bottom of the U, then he slightly arched his back and did an enormous kick with his fluke. That kick thrust his body forward and he slid through the water cleanly with a circle of tiny waves surrounding his upper body. His dolphin kick was beautiful and efficient, and he was totally balanced in the water.

He swam the most beautiful butterfly I had ever seen, but instead of pulling his flippers up over his head, which he wasn't built to do, he kept his pectoral flippers by his sides, using them for steering and turning. He deepened the sideways U by diving deeper with the thrust of his fluke. With a stronger push of his fluke he could dive faster and deeper. He danced with the water and then at times seemed to be part of it.

Taking a breath, I dove under the water and watched him coming toward me. He knew precisely how to move his fluke, balance his body, rotate, and stretch his body and breathe. He had great flexibility and a natural feeling for the water. He knew how to use his flippers, how to hold them out to gain more lift or drag and how to steer. He was the greatest and the most beautiful swimmer I had ever seen. And in only four months he knew more about the ocean world than I would ever know in a lifetime.

But I watched him carefully and attempted to do what he did. Just because I was human didn't mean I couldn't learn something from him. If I simply watched and tried to do what he did, I would learn.

Steve leaned over the railing and shouted to me that we had to try something different. This swimming back and forth along the pier and shore was wasting precious time. The longer the baby whale was separated from his mother, the less chance he had of surviving. If we didn't find her, either he would starve to death or, without her protection, he could easily be eaten by a white shark or a killer whale.

five

Steve devised a plan. He spoke with the local boat operators and fishermen and asked them if they had seen a lone mature female whale. None of them had, but they said they would get on their radios and check with friends working along the coastal waters.

We decided I should return to the jetty, to the place where the baby whale first started swimming with me. We thought the jetty area could be the place where he lost his mother.

"Okay, let's go for a swim, baby whale," I said, and I started swimming, hoping he would follow.

He swam with me for a bit, but he really just wanted to play. He rolled onto his stomach, moved his tail fin up and down a couple times, then flew past me like a rocket. He turned, swam back, rolled on his side, and dove underwater. I just kept swimming toward the jetty, sensing the urgency now more than before. Swimming with my head up like a water polo player, I scanned the ocean's surface.

The California land mass was quickly warming, and the sea breeze was gaining strength, crumpling the ocean's surface into a mass of waves as reflective as aluminum foil. White sunlight caught the edges of the tiny curling waves, and they sparkled like white stars on a sea of blue. The light was so bright it was blinding.

Squinting, I stared across the water, looking for a large spout, a ruffle of white waves, a hole, or a long wide groove caused by the mother whale swimming through the water.

I couldn't see her and I felt a sudden tightness in my heart. What could I do?

The baby whale was swimming right beside me, so full of energy, speeding ahead, circling back, and bounding through the water with great exuberance. He turned and swam right under me, so close I could almost reach out and touch him.

He seemed happy. He seemed to be playing and full of energy, but how long could that last? When had he last eaten? How long would his body fat sustain him? How often did he need to drink so he didn't get dehydrated? What would we do if we couldn't find his mother?

I didn't have any answers. But as long as we tried, as long as we kept looking, there was a chance we could find her.

Sometimes it's the process of doing that makes things clear. If we don't start, we never know what could have been. Sometimes the answers we find while searching are better or more creative than anything we could ever have imagined before.

My mind continued to roam, spin, and focus as I swam. Was there a better way to search for the mother whale? What would happen if she had already moved north? Had she been calling him? Had he missed her? Had he heard her voice for the very last time? Had the baby whale learned how to communicate? Had he been calling her? Maybe she couldn't hear him?

I heard the deep rumble of the Long Beach Lifeguard boat. They were zooming toward us, their bright red bow cutting a V in the gray-blue water. White waves shot up against the red hull, and the V spread into fast-moving waves.

The lifeguards patrolled the coastal waters all year round. Many of them were my friends. Some of them had accompanied me on channel swims and others watched me during my daily workouts. I was always happy to see them, but on this morning I was thrilled. They would be able to help. And they would do their best.

They took great pride in finding and rescuing people, and they also had a deep appreciation for the wildlife that inhabited the ocean and shores. They studied the birds, marine mammals, and fishes and swapped stories about their findings. They observed the behavior of the marine animals and knew all about the migration of gray whales.

Two lifeguards came out on deck. They were older guards, in their forties or fifties, with hard bodies, bronzed skin, broad shoulders, and very big smiles. One lifeguard had fine wavy blond hair and spoke quickly and softly. His friend was taller, with straight dark brown hair. He had a deep booming voice. I recognized them: They had been partners on the boat for a long time.

When they pulled alongside me, I explained that I was looking for a gray whale.

They told me they had been watching the northern migration. That morning they had seen a pod of five whales swimming along the edge of the Long Beach breakwater. The lifeguards had watched them swim outside Los Angeles Harbor, keeping a distance from cargo ships and tankers entering the harbor. Once they crossed the harbor entrance, they made a straight line for the Palos Verdes Peninsula, using it as a land-mark, and then they followed the coast north toward Alaska.

The baby whale surfaced near their boat. The life-guards silently watched in complete amazement. They'd never been that close to a baby gray whale. They'd heard that people swam with them in the lagoons off Mexico and that they enjoyed being touched by people, but they couldn't believe that they had a baby swimming right around their boat.

I started to worry when one lifeguard said that gray whales are very protective of their young. Gray whale mothers usually don't let their babies out of their sight. There was a good possibility the baby's mother was dead. For a moment, I felt great sadness. I don't know why, but I didn't think she was dead. I reasoned with myself: Wouldn't someone have seen a whale floating in the water, especially in an area where there's a lot of boat traffic? Wouldn't someone have seen something? No, she had to be alive.

The baby whale swam ten feet from me. He spouted, bounded through smooth waters, and weaved from side to side. He swam under me and rolled over; I think he was trying to play with me. Whales love to play. They nudge each other, and as they swim their bodies sometimes touch. Sometimes baby grays ride along on their mothers' backs.

I still wanted to touch him to somehow reassure him, but I couldn't reach far enough. So I thought as strongly as I could: Don't worry, little guy, we'll find her.

I glanced across the ocean's surface. It looked so big. I would take you to your home if I could, I thought, as loudly as I could. But you don't really have a home. It's all that water that stretches out around us.

Or maybe it's not that at all. Maybe your home, everyone's home is simply where your loved ones are. Then that's really home. I will take you home. We will find her.

The lifeguards said they would search for her. They would radio Steve if they saw anything at all. They would radio the crews who were on patrol in different waters off Long Beach, and they would radio other lifeguard departments at Surfside, Sunset, and Huntington beaches to see if they had sighted a single female whale, one that seemed lost, or confused, or wasn't migrating north.

So the baby whale and I continued swimming to the north jetty and once we reached the rocks, we stopped and scanned the water.

There wasn't a huge hole of dark blue made by the impression of his mother's body diving into the sea, or a long trail of blue, or even a fluke. There was no sign of his mother.

I treaded water and the baby whale floated near me. We watched and waited for five minutes, and when I didn't see anything, I decided to turn around and return to the pier. But when I started swimming free-style, the baby whale didn't follow.

He dove suddenly and disappeared.

Three minutes passed and there was no sign of him.

I was worried; I knew he could hold his breath for five minutes, maybe more, but he had just vanished.

Taking a deep breath, I dove under the water to look for him. Using wide breaststroke pulls I dug deeper into the water column, moved down through the thermocline, and the farther I went, the colder the water became. The color faded away. Wave surge churned the muck on the seafloor and the visibility dropped to ten feet.

Slowly, with my legs extended over my head, I turned in a circle, trying to see the baby whale.

My ears were filled with the sounds of the city. It was like walking through Times Square in New York City during the height of rush hour. Sound in water travels four times faster than it does in air and it travels farther, but the sounds surrounding me seemed to be amplified by the water. And I felt the energy waves of the fire truck and ambulance sirens, the rumble of a jet airplane taking off from Long Beach Airport, the deep throb of a helicopter flying directly overhead, the sound of swarming mosquitoes as three racing Jet Skis buzzed within a few yards of me.

There were more sounds, sounds I couldn't distinguish, and perhaps sounds that were too low for me to hear. There were so many ships heading for Los Angeles Harbor or traveling north or south along the California coast; maybe their sonar made so much noise that when the mother whale called her baby, he couldn't hear her. Or maybe he was crying out for her and she couldn't hear his cries. Maybe she was using her sonar to try to locate him, but with all the interference from the ships' sonar and other sound waves moving through the water column, she couldn't locate him.

My lungs were burning, so I used large breaststroke pulls to return to the surface, releasing bubbles from my mouth and letting the water lift me back to the surface. I was disappointed. I had thought I would be able to find the baby whale. Maybe he had given up on me. Or maybe he was swimming far away in deeper water.

I decided to swim out parallel to the jetty. Taking three large breaths I hyperventilated so I would have some extra air in my lungs. I kicked my heels over my head like I was doing a handstand on land and made large rounded breaststroke pulls so I could dive down quickly.

As I pulled myself deeper into the water, the waves slowly bounced me up and down. It was like entering a mermaid's world where color and light were transformed into liquid. I swam through colors, through liquid silvers, whites, yellows, greens, purples, and blues. It was like diving into bubbly white champagne, into clear gin, deeper into swaying walls of yellow chardonnay. The water grew colder and colder the deeper I dove. I passed into a world of shimmering julep green, through merlots and grape into heavier waters, and finally into deep water the color of blueberry juice.

Sunlight became liquid too. Undulating beams of white and gold and silver light whorled and wavered around me as if I was lying in the center of a neon hula hoop. The deeper I dove the tighter my goggles felt, like they were squeezing into my eye sockets, and my head felt as if it was clamped in a vise and the pressure was twisting down on my head. My ears and sinuses hurt. Swallowing hard, I released the pressure and pulled past a large field of kelp that was rolling in and out with the surges, riding a liquid breeze. I dove deeper, until the liquid color faded into a soft baby blue.

Using my hands like pectoral fins, making sculling motions like a goldfish, I turned slowly around in a circle. My blood was pulsing in my temples. My body felt squished in by the water pressure.

Stay under just one more moment, I told myself. Look again. Turning around in the mermaid world, I strained to see something, but there was no sign of the baby whale.

My lungs were on fire from my desperation to breathe, my head throbbing from the buildup in my blood of carbon dioxide that I couldn't exhale—it increased the blood flow to my brain and my urgent drive to breathe. I shot back to the surface, hoping I could reach it before I ran out of air, before I blacked out from the lack of oxygen, and drowned. My ears were popping from the acute changes in pressure, my muscles ached from lack of oxygen, and my lungs felt like they were imploding and screeching for life.

Three bubbles, two bubbles, one bubble—hold on to that last bubble. Intense pressure. Pain. One last huge silver bubble flew to the surface and burst.

Gasping for breath, lying on my back, treading water, body one big ache, I breathed hard and fast. My heart was pounding and my head felt like it was about to explode. I lay there floating, trying to get my breath back to normal. Trying to get my heart rate down. Trying to get back in balance.

One more time, I would try. But I knew I needed to rest for a few minutes. I needed to get the lactic acid out and more oxygen into my blood.

Lifting my feet up, remembering to pretend that I had a quarter between my shoulder blades that I needed to press into the water, I floated, letting my mind relax to take the pressure out of my head. Looking up into a sky filled with the colors of Provence, the bright blue of the sky, the yellow of the sun, and the white of the clouds, I watched the clouds become beluga whales, angelfish, sticky macaroons, woolly llamas playing flutes, the summit of Mont Blanc, great cutter ships and white elephants and fluffy white house cats, furry borzois and the great Pyrenees. Floating like the clouds I rode the ocean currents.

The water suddenly became gritty, discolored, earthy brown and red, flecked with shiny mica. I was in a small river of sand and silt, caught in a riptide flowing offshore. It was a small riptide, moving at only one knot or two, but in a few minutes, it had carried me two hundred yards offshore. I knew it was nothing to be afraid of; it's only when you try to turn and swim directly into a rip that you have problems. If I needed to get into shore at any point, I could get out of the rip by just swimming fifty or a hundred yards parallel to shore, and then I could swim to shore. I enjoyed the free ride on the riptide into deeper water, and the trip gave me inspiration for a new plan.

This time I would dive deeper and faster with the hope of finding the baby whale. Taking seven deep breaths, I dove and pulled as fast as my arms would move, and pulled myself down twenty-five feet. A school of bat rays were swimming in single file, flapping their wings, swimming right toward me. They grew bigger as they swam closer. From wing tip to wing tip they were five feet wide, and they must have weighed two hundred pounds each, maybe more. Their bodies were flat, their skin was smooth and like a surfer's wet suit. Their heads were large and protruded in front of their fins.

One bat ray swam within three feet of me. He moved gracefully, like an enormous newspaper rolling and unrolling. Six more bat rays followed in complete synchrony. They swam to a level ten feet above my head, revealing their white underbodies. They turned in toward shore where the water was warmer and climbed into the upper inches of water, skimming the light blue surface silhouetted by the rays of the sun, their long whiplike tails trailing behind.

They looped back and dove, flying past me, one after the other as they flapped their fins fast, using the downward thrusts to move the silt on the ocean floor and uncover other stingrays and halibut hiding from predators. The bottom erupted as stingrays scurried by and bat rays closed the distance. Not wanting to leave, but needing to breathe, I turned and followed what was left of my bubble stream to the surface, trying to push the water downward rapidly with my arms the way the bat rays did.

Snapping through the water's taut surface, I rolled over onto my back and breathed fast and deeply.

I didn't stay there long. In the back of my mind I knew that if I didn't find the baby whale soon, he might never find his mother, and even if he was with me, we might not find her. But still, there might be a way to help him.

If I continued diving in the same place, over and over again, did I increase or decrease my chances of finding the baby whale? I wasn't sure, but I decided to try something different. This time I swam two hundred yards farther offshore into a warm current. Taking three deep breaths, I bent in half and lifted my legs up over my head, pulled rapidly, and felt the water squeezing around my head like a vise that continued to tighten as I dove deeper. My heart was beating slowly in my throat. I pulled a few more strokes, and then sculled to hold myself in place.

Two green sea turtles swam off to my right side. They were large, their carapaces mottled in patterns of browns, greens, blacks, and grays. They were about four feet long and could have weighed one hundred and fifty pounds. They had to be at least fifteen years old. And they were swimming slowly and easily. Their fore flippers were beating the water like wings and they weren't in any hurry, carrying their homes along with them like aquatic RVs. They were amazing animals, able to hold their breath underwater for up to five hours, and they could make their hearts slow down so they beat at nine beats per minute. As they disappeared from view, I pulled back to the surface with a deep sense of gloom. There was no sign of the baby whale or his mother. What could I do now?

Floating on my back trying to catch my breath and energy, choppy water rolling over my shoulders and arms, I stretched them out and let them float near my head, tucked my chin to my chest to stretch the back of my neck, and then grabbed my knees with both hands and pulled them into my chest one at a time. Slowly I released them. That took the stress out of my back and I imagined that I was in a giant cradle rocking from side to side, with gentle waves rolling under me and massaging my back and shoulders.

The wind blew through the funneling waves,

transforming them into wind instruments. They were giant bassoons, tubas, trombones, piccolos, trumpets, flutes, clarinets, saxophones, French horns, B horns, C horns, and oboes. A symphony of the sea was playing. And the music the waves played grew louder, changed in tone, in pitch, and in length with their constantly changing shape and the amount of wind blowing through them.

As the waves broke, a new movement emerged. The rush of the waves plucked the beach like the strings of a harp, making high, sweet notes. The swirling breezes strummed the water's surface and the tiny wind waves sounded like the flowing notes of a piano.

In the background the deep resonant clangs of the shipping buoys, plaintive cries of the seagulls, and calls of the willets became a part of this great sea symphony, and I enjoyed each movement until I heard an incessant high-pitched whine, like the outboard engine on my grandfather's fishing boat.

Quickly, I lifted my head and moved my arms fast, splashing them hard against the water to make myself more visible.

I could tell by the way the skipper was holding the

motor and peering over the bow of his boat that it was Carl. My mood immediately lifted. I raised my arm high above my head and waved. Carl was an old man who fished along the shores of Seal Beach and Long Beach. He used worms like my grandfather and sometimes lures, but he believed he caught the most fish with night crawlers. He had a compost pile at the back of his house where he tossed his grass clippings and vegetable peels, and he grew the longest night crawlers I'd ever seen.

Carl tipped his white sailor's cap and turned his boat toward me. His face was red and deeply etched from years of being on the water. He wore dark sunglasses to protect his light blue eyes.

Carl loved to stop and talk. His wife had passed away long ago and he still missed her company. But while he liked people, he enjoyed fishing alone; that way he didn't have to wait for anyone or be on anyone else's schedule. Sometimes he brought an older friend—he was kind of a grouch—but Carl did it because he said his friend's wife had passed away too and he was lonely. I thought it was nice of Carl to do that, since I'd much rather be alone than be with someone cranky. Carl and I didn't see each other very often. It usually took him some time to get himself going in the morning; he was usually starting to fish when I was finishing up my workout.

But I loved to see him. He always had some news or information that I could think about when I swam. Just as good, Carl usually caught an extra halibut or two, and he always gave me some to bring home for dinner. It was always a little strange kicking ashore while holding a five- to ten-pound dead halibut above my head with fish juices sliding down my arms.

No fish ever tasted as fresh or as sweet as the ones Carl gave me. I liked the fish even more because they were from Carl and I could tell he was as excited about giving me the fish as I was receiving them from him.

Carl was perplexed. He saw me floating on my back: He had never seen me do that; usually I was swimming on pace. He thought something was wrong, and when I told him about the lost baby whale, he smiled as if he had just been given an answer to another of life's mysteries. He hadn't had a nibble on his line all morning and it hadn't made sense. The fishing conditions seemed perfect, and usually he caught at least two or three fish.

The baby whale must have scared the fish away. Carl thought the baby was still somewhere nearby. He told me he would cruise the shore and radio Steve if he saw anything at all. Sometimes, he said, the important things take time, sometimes they don't happen all at once, sometimes answers come out of time and struggle, and learning. Sometimes you just have to try again in a different way.

He knew so much more than I did, and I always liked talking with him. He turned his boat, glanced back over his shoulder, tipped his hat, and motored along the shore.

Try again one more time. Try diving into deeper water. If you can't find the baby whale this time, then it's time to try something else.

Diving below the water, I pulled as fast and as hard as I could to get down as deep as I could go. From moment to moment the world changed. I swam through a brilliant melting kaleidoscope of green, yellow, indigo, violet, and soft blue. As I pulled deeper, the pressure tightened around my head and body like an invisible shrinking knot.

My ears popped, and I pulled deeper. The increased arm movement was using oxygen more quickly, so I had to keep enough air in my lungs so I could make it to the surface without passing out. I was grateful for those training sessions my coach had given me where I had swum one lap of the pool breathing every five, then seven, nine, and eleven strokes. Still, I wasn't used to swimming with the weight of the water on me and I knew I couldn't stay down too long.

There he was floating right below me, inches above the soft light brown silt-covered bottom. He looked right at me with his large brown eyes. He was so peaceful. I laughed and I wanted to swim over and hug him. He had been there all along, just watching me.

He lifted his fluke, did a little nod downward with his head, and glided toward me underwater through beams of white and green sunlight. The liquid light dappled and waved along his skin.

He swam in a small circle and I laughed out of relief and delight. And tried not to get a noseful of water. He wanted to play with me, but I was out of air.

Following my bubble stream, I raced to the surface. My lungs were down to zero.

Floating on my back, I gasped for air, caught my breath, and then dove again.

He had the most incredible set of lungs. He was able, it seemed, to stay down forever. He did one giant dolphin kick and slipped through honey-colored sunbeams, and they changed with his movement through the water, becoming squiggles of lime green light.

He grunted softly, squeaked, paused, then grunted softly again. He paused longer this time, as if he was waiting for me to respond. Then he clicked and chirped. He made a small symphony of underwater sounds: high and low tones, soft and loud; all were new and different.

For the first time in my life, I heard a baby whale speak. I heard the voice of the whale. I was thrilled. The baby whale had spoken to me.

More than anything, I wanted to talk to him and I wished I could understand what he was saying. It was like going to a foreign country and not being able to speak the language. It was frustrating, wanting to somehow make a connection but not being able to understand anything.

He looked at me with his big chestnut brown eyes and I wanted to reach out and touch him. I wanted to be able to do something that he would understand.

Instead, I just watched him, trying to think of a way I could help. Watching for any gesture he might make, anything I could comprehend.

He tried a few more sounds, louder in volume, higher in tone. And he waited as if he was expecting me to say something.

When I didn't say anything, he turned on his side and looked at me. He opened his mouth. He had a large pink tongue. I think he was clicking it against the roof and base of his mouth, like a human child. He continued talking or vocalizing.

And I listened to the sounds with real awe. Years later, I realized that if I had found the baby whale on my first dive, I might never have heard him speak underwater, I might never have seen the graceful bat rays or the swimming sea turtles, and I never would have known how far I could go down into the ocean depths on a single breath.

six

There was no sign of the whale's mother by the jetty, underwater, or anywhere else so I started swimming back toward the pier, hoping the baby whale would follow. He didn't.

I thought that if I could communicate with him he would come with me, like a dog responding to a familiar whistle. I thought that maybe if I could try to speak in his language he would understand. I tried to repeat his chirp. It was pathetic. It didn't sound anything like him. I tried to grunt, a really big grunt, but all I got was a noseful of saltwater and tears in my goggles from the salty sting. I returned to the surface to clear the water out of my nose.

And it finally occurred to me: No matter what I sounded like, I didn't know what his sounds meant, and even if I could imitate them, I wasn't going to be anything more to him than his echo.

Unable to figure out a different approach, I resumed my swim back to the pier.

Sometimes it makes sense to try something again and keep it simple. A moment later, the baby whale took the lead.

When we reached the pier, Steve was waiting there along with a group of fishermen and a handful of locals and tourists. Steve said that one of the fishermen on an offshore boat thought he had sighted the mother whale near one of the oil rigs.

The oil rig was about a mile and a half offshore and it was almost in a direct line with the pier. I had swum out there only once before, during an open-water race, but at that time, I had had a paddler with me on a long paddleboard. He had helped me stay on course, and he had watched for danger.

But the baby whale had already turned and started to head offshore. He looked over at me as if to say, Please come swim with me.

I knew it made no sense to follow him. I could think of many reasons why I couldn't or shouldn't, but I didn't want him to go off alone.

Sometimes things just don't make sense, sometimes there's no reason to explain how or why I wanted to do them; I only knew that I had to, I had to try. Without trying I would never know what could happen. It was like reading a great mystery and never knowing how it finished, always wondering who did it. Sometimes the things that make the least sense to other people are the ones that make the most sense to me.

Maybe I knew this, too, because I didn't always fit in. I was shy and large, and I believed that I had to work hard and study hard to do well. I had different friends—from computer wizards to the guys on the water polo team and the girls on the swim team to friends in drama and music—but I didn't fit into any one group. I had things I knew I wanted to do and didn't play the teenage boy and girl games. I was more interested in studying people who had been leaders, made discoveries, or explored, men and women who were always going against established thought. It was always difficult to swim against the tide, doing something new or different, because the ideas that could result might cause something to change. Many people are happy with things as they are. They are comfortable with what they already know. But if I didn't move outside my comfort level, how would I ever experience anything new, how would I ever learn, or see or explore? I believe that each of us has a purpose for being here, that we have certain gifts and certain challenges we need to learn from and fulfill for our lives to have meaning and richness.

"I'm going to swim with him," I shouted to Steve.

"I don't like the idea of you being out there alone," he said.

I was afraid. But I knew I had to. Sometimes I just did things because I thought I could and because if I didn't an opportunity to learn something, grow, and reach farther would be lost. There wasn't time for a long discussion. The baby whale was turning out toward the open sea, and I was afraid that if he left now without me, we would never know if he found his mother or what happened to him. Maybe my presence could even make a difference.

So I quickly told Steve I'd be fine and asked him to let his friends on the fishing boats know that we were out there. They'd let other boaters in the area know. He still didn't entirely like the idea. He was an adult and pretty conservative, and he warned me that the closest fishing boats would be a quarter mile away.

"I'll be careful. Besides, I'll be swimming with the gray's son. I'll be swimming with Grayson," I said, and smiled with more confidence than I really felt.

Steve smiled. "Grayson, that fits. He's grace in the water and he's the gray's son."

But then Steve's tone grew suddenly serious, and he advised me: "Lift your head up often and look all the way around you. If a boat approaches, you move out of the way. Don't count on them seeing you."

I swam with Grayson one hundred yards off the pier, two hundred yards, three hundred, four hundred, and on a breath, I looked back over my right shoulder. The pier and the people on it were becoming smaller and smaller. We continued swimming near each other. Grayson led the way. He swam directly toward the oil rig and I followed in his wake. A couple of times he slowed down and stopped dead in the water. He seemed restless and sort of agitated. He probably hadn't eaten for at least a few hours. His energy level had to be dropping.

"Come on, Grayson. Let's swim out there and see if we can find your mother," I said, encouraging him, knowing he couldn't understand a single word, but hoping he would somehow understand the thought.

Words are sometimes too small, too confining, to convey the depth of thought and strength of emotions. How does a whale communicate love, hope, fear, or joy?

He looked so small in the enormous sea and I wanted to protect him somehow.

Maybe you communicate with your heart. That is what connects you to every living thing on earth. Use your heart. It is love that surpasses all borders and barriers. It is as constant and endless as the sea. Speak to him with your heart and he will hear you. No matter how close or far away she is, she loves him. And from that he will have strength. He will.

Let him know that he is also in your heart.

The sky was changing: Thin clouds were masking the sun and the water was becoming a dull opaque blue. The water temperature was dropping too. It must have been about fifty-three degrees.

There were a few fishing boats on the horizon. But as I followed Grayson's "footprints" in the water—the indentations he made with his fluke in the ocean's surface as he swam—I grew increasingly uncomfortable.

Unconsciously, I turned and looked at my feet. The tiny footprints they made when I kicked dissolved instantly. I shuddered.

There weren't any breakwaters or jetties to buffer the strength of the current. Using the oil rig as a reference point, I could tell that we were drifting to the north at about a knot, a little faster than one mile, per hour. The oil rig that had been directly in front of us was sliding to our left. And the ocean's surface was cracking with a northwest breeze. The sea was rising into waves a foot high.

Grayson was swimming hard against the resistance of the waves. He was breathing more rapidly, his *poof*-ing sounds were more frequent. He seemed to be very stressed.

And he was changing course abruptly. He was swimming north toward the oil islands off Long Beach, and then he turned in a half circle, and swam south toward Surfside Beach. It seemed as if he couldn't decide what to do. Then he came to a complete halt.

He hung on the water's surface. His eyes opened wider than before.

"What is it, Grayson?"

He turned toward me, and he tilted his head and looked at me with one eye.

He seemed to be waiting for me to follow him.

I really didn't like being so far from shore. But I swam toward Grayson anyway, with my head up.

There was something in the distance, floating on the water's surface.

We moved closer. It looked like white lily pads were floating on the water.

We swam nearer and the lily pads grew larger. They were ovals three to four feet in diameter with scallop-shaped tails. The ovals were different colors—gray, olive, black—and they fluttered.

They were giant fish, giant ocean sunfish called Mola mola, basking on the ocean's surface, absorbing the sun's warmth through their skin. They shimmered silver, and as the light shifted they became luminous and ivory like the moon on a clear black night. They had small dark eyes and light pink oval mouths attached to a snout. They were the heaviest bony fish in the world, weighing up to five thousand pounds.

One sunfish was swimming. He was waving his top fin and bottom fin, using his pectoral fins as stabilizers and his tail fin as a rudder. And he was spitting water out of his mouth to help steer.

He dove deeper and deeper and deeper into a cold current to cool off, and when he resurfaced, he rolled over to let the sun warm the other side of his body.

Grayson maneuvered between the shimmering sunfish; they seemed oblivious to our intrusion. And we continued heading toward the oil rig. I felt very exposed; my legs were dangling like worms in the water.

Four hundred yards from the oil rig's base, we entered a sea garden. It was filled with long ribbons of golden brown kelp, which had short ruffles and a mermaid's necklace of pearl-shaped air bladders attached to the main stem that enabled the kelp to float and dance on the water, signaling the speed and direction of the water currents.

On the seaward side of the oil rig, a large cluster of kelp smoothed the waves and we were able to swim to within two hundred yards of the rig.

The oil rig rose above our heads like a mini–Eiffel Tower with metal cranes and drilling equipment that towered twenty feet or more above our heads. These were connected to a large metal platform and the platform was attached to multiple metal stilts that had been drilled deep into the ocean floor.

The oil rig was an amazing and yet ominous structure. As the rig pumped oil out of the ocean floor, I could feel its energy emanating through the water. It felt very different from the natural energy of radiant sunshine or the quiet energy of the earth.

It felt like being in New York City. Being among the city's skyscrapers was like standing between power transformers with the energy flowing all around all at once. All of this energy bounced off the surfaces of the buildings and was amplified by the wind blowing through the open spaces. The energy from the oil rig was like that, but it was more diffuse, a softer force that was transmitted in waves through the water.

The energy from the oil rig was a constant hum—a sort of *ooommm*. And there were loud metallic noises, creaking, groaning, clanging, and hammering.

Men who worked on the oil rigs had told me that they noticed the energy attracted fish into the area and lulled them to a state of inactivity. There was a deep-water-fish metropolis around the oil island.

As I breaststroked closer to it, I noticed schools of sunfish clustered together near the base of the rig, floating peacefully on the water's surface, their bodies conforming to the shape of the waves rolling under them.

Grayson swam past the sunfish, and he didn't even notice the green sea turtles paddling by, like a green turtle swim team. They all pushed off near the oil rig and swam together as if they were setting off on a series of sprints.

Slowly, a school of sea bass swam past, moving like a shimmering curtain of silver blue.

Grayson took a big breath and dove five feet down, past a cluster of clear moon jellies. They were beautifully transparent except for white circles on top of their domes. Grayson swam by purple jellyfish that were larger, like large Jell-O salad molds, and they were beautiful, graceful swimmers. They moved by contracting and expanding their domes, like opening and closing umbrellas.

Their long, flowing tentacles stretched up to six feet beneath them. I hoped they would stay below me. The moon jellyfish didn't sting, but the purple ones did. The purple ones had tentacles that had tiny little barbs attached to them. The barbs were trigger-loaded with stinging cells called nematocysts. When a swimmer brushed up against a tentacle, the barbs snapped off or stuck to the swimmer and that movement fired the stinging cells. I had been stung before and it hurt. The intensity of the sting depended on the type of jellyfish. The sting of the Pacific jellyfish wasn't as bad as a bee sting, but a swimmer could be stung multiple times at the same time. It felt like running through a field of nettles naked.

Grayson knew to avoid the tentacles. Diving into the deep water he wove his way down through the sea of purple jellyfish and out of reach of their tentacles.

Swimming on the surface, a pair of bright orange garibaldi greeted me. Garibaldi were fish that resembled giant goldfish. Usually they inhabited the shallow coastal waters along rocky shores with lush kelp beds, where they could hide from predators behind the veils of kelp. Seeing them swimming so far offshore was very uncommon. But they were a protected species and there were many garibaldi living off the shores of Laguna Beach, and also along the shores of Catalina Island. They were attracted to bright orange or tangerine colors, and whenever I wore a tangerine swimming cap, they swam around my head. It wasn't until a friend pointed it out to me that I noticed the garibaldi had sharp front teeth, which they used to crack open the soft shells of spiny sea urchins. Then they popped the round orange roe balls out of the shell and ate them whole.

This pair of bright orange garibaldi seemed to be mates. They swam side by side very close to each other through the long, slowly waving tendrils of brown kelp. And they swam around my head, checking to see if I was a garibaldi invading their territory. They seemed satisfied that I wasn't. They swam to within a few inches of the oil rig. I watched them become two orange dots in the dark blue sea.

Grayson continued his dive, deeper and deeper into the enormous sea, and I watched him. Why are you going so far down? I felt myself getting a little nervous. How long can you hold your breath?

Be careful, Grayson. Be really careful.

Grayson's fluke became a tiny waving gray Y in the light blue depths and then the Y disappeared into the darkness. He dove so far down, one hundred or two hundred feet, that I wondered how he could stand the pressure changes in his ears and head. How could he equalize that pressure so quickly? How come his ears didn't rupture? Would he have enough air to return to the surface?

I knew Grayson was born to swim and dive to great depths but I still held my breath with him. Unconsciously I always did this when I was teaching people how to swim. I never wanted them to run out of air on the bottom of the pool and have a bad experience. I held them by one arm and pushed them back up if I thought they were going to run out of air. I took another breath and ran out of air again. I took another breath and repeated taking breaths twenty more times. He was still gone. I looked at my watch. He had been underwater for at least five minutes.

Was he okay? Would he return? Where was he?

The sun shifted suddenly, highlighting the water below so that it was possible to see down into the deep.

I said to myself, This is the dumbest thing I've ever done, swimming so far out without a boat. Then I thought, No, I've done dumber things, like the time I was five years old and decided to climb down into a rock quarry, alone, so I could see the shiny red cranberries floating in a natural pool way down at the bottom of the bog. It was dumb when I tripped on a rock, slid down a cliff, and nearly fell into the quarry, but luckily I caught a tree branch and hung on until my mother found me. Yep. That was dumb. It was almost as dumb as the time my brother convinced me to jump out of the Hatches' barn window to test if the snow was soft or hard before he and his friends followed. It was hard. And it really hurt. It was dumb, but I did something dumber than that when I was seven and I kissed Craig McQuade. That was really dumb.

Yep, but that wasn't the dumbest thing you've done: Remember when you and Sue and Kari and Kittridge had a sleepover and you toilet-papered John Mill's house? And he caught you? Remember the time you went ice-skating on a pond after you'd been told the ice was too thin, and it was? Remember the time you said the S-word and your grandmother heard you? Hmm, remember the day you forgot you were going to have a math quiz? Remember the time you stayed out too late and got lost in the woods? Remember last month?

Okay, okay, I'd done a lot of really dumb things, but this definitely had to be *one* of the dumbest.

Making a few arm strokes, I took a breath and looked down. It was dumb to look into an abyss. I have no idea why I kept doing it. I guess I just wanted to see more. I wanted to understand what I didn't already. I was just curious. I couldn't help myself.

The water was navy blue and full of wavering and shifting shadows. I moved a little deeper into the shadow of the oil rig. Taking another breath, I looked down again. I couldn't see Grayson.

The sea seemed empty without him.

Suddenly I felt more alone than ever before. I was scared for me and for him.

Could he dive to five hundred feet like an adult whale? How long would that take? How long could he hold his breath? Where could he have gone? Would he return?

I wondered: How long do you wait? How long do you wait for anyone?

I hung on the upper inches of the water and wondered.

Should I go?

No, he has come to me for help. I have to find him.

I pulled myself underwater with wide breaststroke pulls. I dove deeper and deeper into the dark nothingness.

My head throbbed with the pressure. The dark blue world whirled around me. The emptiness tightened around me like a boa constrictor. I waited. I held my breath until I was nearly out of air then raced for the surface.

From the tension of the dark empty depths an idea began to emerge.

It was that space between not knowing and knowing, that tension between losing and finding, that blank page between silence and song, that emptiness that creates the need to create, to try, to imagine, to solve.

I stared down again and it was like looking straight down from the top of the Empire State Building, but I couldn't see the ground or Grayson, so I swam farther out, floated on my stomach, and looked down again.

The sun shifted slightly and the water became more transparent. Now it was like standing on the very edge of the Grand Canyon and looking down deeper and deeper as if a trapdoor had opened below and the bottom was dropping. (Now the water was so clear I was afraid it wouldn't support my weight.) I felt like I might fall into forever. I had to lift my head up quickly and take a couple of fast breaths.

I don't know why, but I had to look down again. I had to see one more time if I could find Grayson. I had to figure out how to look down without feeling as if I was falling.

I imagined I was hovering a foot above the moon's surface looking down onto the earth and ocean. When I visualized myself way above the earth, the water depth no longer seemed so frightening. By changing my thoughts, I was able to alter my perspective, to calm down, and to refocus.

I floated on the surface, rolled onto my back, and told myself to relax, and as my body relaxed, my mind did as well, and then new ideas began to flow.

Maybe I couldn't help Grayson, but I knew that he had someone out there with him, and sometimes just having someone with you is enough.

There are all sorts of ways to think about the world, and so many people who think differently. Still, I believe there are two basic ways of thinking: one of possibility and hope, the other of doubt and impossibility. When I think about impossible things, I think of a friend of mine who did the impossible, and that makes me believe impossible things can become possible. If I try, if I believe, if I work toward something, and if I can convince other people to help, the impossible isn't impossible at all.

At age fifteen, I swam across the English Channel. Most people didn't think that was possible. And now, looking back through time, I remember that my friend Greg Miller once asked me to drive with him at three in the morning to a deserted runway in Bakersfield to watch him attempt to be the first person to achieve human-powered flight. His goal was to fly the *Gossamer Condor* a mile in a figure-eight course. The plane was created by Paul MacCready and a group of friends from Caltech; its wings were made of balsa wood and Mylar sheeting, and were attached to a bicycle with piano string.

All Greg had to do was to pedal as fast as he could to get the airplane airborne, keep pedaling as fast as he could, maintain an altitude of twenty feet off the ground, and keep the plane stable while making a giant figure eight.

Greg became one of the first to achieve human-powered flight. Like him and his crew, I simply believed he could do it. He worked hard, trained hard, studied birds in flight and in taking off and landing; he looked to nature and tried to emulate it. And he worked with a group of rocket scientists and inventors who saw the possibilities in life; they saw what was and what could be.

I never questioned why Greg would want to attempt to be the first person to fly using human power. It was something that he and his team believed was worth achieving. And I think I felt the same way about the baby whale. Like Greg I believed in trying to do things that people may have thought impossible.

I believed we would find Grayson's mother in the vast ocean.

I had heard that whales sometimes dive into very deep waters so they can talk to each other. Their voices carry a much greater distance in the deep, where the water is denser and colder.

I wondered if that was why Grayson had swum so far down, to listen or talk. He had been gone for seventeen minutes. And it had seemed like forever.

I put my face down in the water, and in my mind I shouted, Grayson! I hoped he would somehow hear me.

Grayson popped up from below the water and swam beside me.

"Grayson, you are so beautiful. How in the world did you ever find me? How come you came to me to help you? How could you have known that I would?"

Grayson rolled over and looked at me. In his eyes, I

saw a brightness, a sense of vitality, and a gentle sweetness. I held him in my eyes and in my heart.

His poor mother, though, had to be frantically searching for him. How in this big ocean would she ever find him?

Do what you can do, I thought, don't get overwhelmed by the enormity of something. Break it down into smaller pieces like you do when you swim. Do one thing at a time.

"Grayson, let's swim back to shore now," I said. I had to. I was cold. And tired and depleted. My eyes were burning from the saltwater leaking into my goggles.

Grayson seemed to understand. He turned with me and started swimming toward shore.

The current seemed to rise on our backs as if a giant hand was lifting us and carrying us toward shore. I felt a deep sense of relief. I was ready to reach the beach.

But all of sudden, Grayson dramatically changed course.

He turned almost completely around. Had he heard his mother's voice?

seven

Grayson was swimming so fast underwater that I could see streams of darker water flowing over his head.

He was dolphin-fast; his footprints were snapping up to the surface with each giant flick of his fluke. His footprints were spaced closely together.

Grayson was holding on tightly to the water. As his speed increased, the resistance increased. He held on tighter and tighter, swimming faster and faster, and he moved through the water in a line as straight as a torpedo.

Then suddenly he leaped out of the sea. He

transformed his body into something like a giant cymbal, with the sea's surface as the other cymbal. He breached.

He intentionally hit the water at a sharp angle to make an enormous impact that would create noise— and a giant splash. It was the best sideways cannonball I had ever seen.

Gray whales often breach to dislodge barnacles and sea lice from their skin. But they also do it to communicate with other whales. And I hoped he was trying to communicate with the other whales that might be passing through the area.

He paused and then took off again, moving like a plane speeding down the runway for takeoff. He converted his speed across the water into lift and, this time, got his body more airborne. He flew five feet above the sea's surface and when he hit the water I felt the impact. His splash soared six feet into the air.

For a couple of minutes he caught his breath and then he sped off one more time, faster than before, and this time when he leaped, he adjusted his position in midair and launched himself ten feet across the water. This time the splash flew twelve feet into the air. If there had been an ocean Olympics for the long jump, Grayson would have received a gold medal.

I waited for him. And wondered how much longer I could wait. I was cold and tired. But, I thought, maybe he's down below talking with the other whales that are traveling along the whale trail north. Maybe they heard his breaching sounds. And maybe they know where his mother is or maybe they will tell her where to find Grayson.

All that I knew, though, was that I was really tired. And that soon I would have to get ashore. I strained my eyes to spot Grayson's fluke. Pulling my goggles off my face, I placed them on top of my head and rubbed the soft skin where the goggles' rubber gasket had pressed against my eye sockets and forehead. They were sore and my eyes burned from the salt-water. My neck hurt from lifting it way too much. I was whining. I checked my watch. It was nearly nine a.m. Grayson had been gone for ten minutes. I knew I should head home soon or my parents would be wondering about me.

I made my feet move in small circles, toward each other, one foot and then the other, like an egg beater, then I did it faster so that the action would lift me two or three feet above the water and I could get up higher in the water to see if there was anything out there.

It was hard work. I was breathing heavily so I decided to alternate between doing a slow and a fast eggbeater. I rested during the slow eggbeater and worked on the fast, and I moved slowly in a tight circle so I could see all the way around. Twelve minutes had passed since I had last seen Grayson.

Putting on my goggles and taking a breath, I put my face under the water, and made small slow circles with my feet.

Had he found his mother? Or had he swum off? Should I swim to shore alone?

Shore was farther away than it had been ten minutes ago. The ebb tide was pushing me beyond the oil rig. Catalina Island's mountain peaks were clearly visible in the distance. I reminded myself that I had swum there before, but that didn't matter. I was becoming impatient.

How much longer do I wait?

I told myself to wait for five minutes, and when five minutes passed, I asked myself again: How much longer should I wait?

The answer came to me. Wait as long as you need to. The waiting is as important as the doing: it's the time you spend training and the rest in between; it's painting the subject and the space in between; it's the reading and the thinking about what you've read; it's the written words, what is said, what is left unsaid, the space between the thoughts on the page, that makes the story, and it's the space between the notes, the intervals between fast and slow, that makes the music. It's the love of being together, the spacing, the tension of being apart, that brings you back together. Just wait, just be patient, he will return.

But the reality was that I was growing colder, more tired, and more hungry.

I checked my watch. Three minutes. It had seemed like thirty. My thoughts were becoming negative.

I knew that if I changed my thoughts, I would change the way I felt about what I was experiencing. I was hungry and all I could think of was food. So I let my mind go wild and I began imagining what I would have when I finished my workout.

All I wanted was a toasted and chewy bagel with peanut butter, or with jam; or a flaky, slightly sweet, buttery croissant and hot, rich French coffee and milk. A cup of hot chocolate with a mound of whipped cream as big as Mount Baldy in the distance. I could eat it and float on it. All I wanted was a pot of hot orange spice tea and a chocolate chip scone. That would be delicious too. Even better, thick moist chocolate cake with chocolate butter-cream icing, or carrot cake with pecans and cinnamon and clove, pineapple, and coconut, or a slice of hot apple strudel—any of these would do.

My stomach was moaning, sputtering, and growling. I couldn't help myself, the more I thought about it, the hungrier I got. I imagined a steamy plate filled with penne pasta and thick marinara with thin shavings of Parmesan cheese, or a dish piled high with linguini and scallops, shrimp, mussels, in a white wine and garlic sauce. Or salmon, I love salmon—grilled, poached, marinated—or New England lobster with butter, or steamed clams. Really anything would be more than fine. I would love a cracker or a thick, juicy grilled New York steak or a rare filet mignon with spinach. Something hot and spicy would warm me up from the inside out: I would love Thai eggplant, Indian jade curry, Hunan beef, Sichuan shrimp, or a hot steamy bowl of sukiyaki. I could eat a cheese-burger with Muenster cheese, or I could eat a Chicago or New York pizza, with mushrooms and long stringy cheese. I kept thinking of food and I got hungrier and hungrier. Then I thought of Grayson. And that made me feel guilty. I had discovered guilt is a great motiva-tor. I thought of him instead of me. And his needs, not mine.

Grayson had to be famished. His mother must be too. She hadn't eaten at all during her migration south, during the birth, or on her migration north. It was amazing the way she swam north with her baby, fed him, giving herself to him with her milk, her body shrinking as his body grew. I hoped we could find her. Sometimes you just had to believe that things would be okay. Sometimes it made no sense to be optimistic, but it sure beat being pessimistic.

The wind was diminishing, the ocean becoming smoother. There were long silky areas and lined sections rippled by the breeze: They made the ocean look like the petals of a flower. As the sun shifted, the water changed to a bright purple blue and it was like floating on Van Gogh's irises and across the fuzzy yellow, gold, and white blaze.

Rolling onto my stomach I took a breath, looked deep into the purple-blue water. Seeing nothing but purple, I closed my eyes and listened. There were so many sounds. Tiny crackling noises like plankton bumping into one another, and the sound of shrimp growing.

Grayson had been gone for fifteen minutes.

But I had been in the water for three and a half hours. And the water was three or four degrees colder out near the oil rig than it was near shore. And when I was floating, I wasn't creating any heat through exercise. The cold was starting to work its way deep into my muscles and I knew that I was getting closer to hypothermia. If the cold water made my body temperature drop too far, I could pass out or die from exposure.

I had to start moving.

I told myself to try one more time. I dove under the water and thought as loudly as I could: Please, Grayson, don't give up on me. Please don't leave me out here. We'll find your mother. I'm not sure how, but we have a better chance if we stick together. Grayson, please come back.

eight

The tide was pressing into me. It was like being teth-
ered to a giant elastic band. I would make some head-
way and then I would be pulled backward. I had to
start swimming faster than I had three and a half
hours ago if I was going to get across the tide and make
it back to shore. I imagined that I was a tiny boat and
my arms were the oars. I pulled harder.

On the horizon were the San Gabriel Mountains.
The range rimmed the Los Angeles Basin and formed
a long arc along the horizon. They were covered with
a light, bright white powdering of snow. The highest
mountain in the chain was Mount Baldy just off to the
right.

As I swam, I focused on Mount Baldy to keep a straight course. Now and then, I turned around and looked behind me and sighted off the peaks on Catalina Island. By imagining that I was drawing a line from the mountains on Santa Catalina Island to Mount Baldy, I was able to maintain a fairly straight course.

I told myself to swim for twenty minutes and then I lifted my head.

I wasn't making much progress. The pier was a mile away. It looked like a long-stem rose held out at arm's length.

The sea surface was changing from a smooth, light silvery blue to a rumpled navy and white.

In the distance the wind was blowing stronger and the sea was becoming increasingly choppy. It felt like I was swimming uphill. And I knew I'd have to seriously watch for windsurfers and sailboats. If I wasn't careful I could easily be run over or get a skeg in my head.

Wind gusted to twenty knots and tossed up more waves. The waves were hitting me in the face. It was hard to breathe, and I could barely see a foot in front of me.

"Grayson, I hope you can hear me. Please come back and swim with me. I need you."

Grayson returned!

He surfaced beside me. He rolled over and floated like a runner who had just finished the last hundred-yard sprint of a marathon. He was *poof*ing and *poof*ing very quickly and deeply, as if he couldn't catch his breath. His body was moving up and down as his lungs filled with air and pulled in more. His body was fighting to recover from a very deep dive.

"Grayson! I am so happy to see you! I was so afraid something had happened to you. But you're back now, you're back now, my dear little friend." I felt such joy that he had returned.

Grayson floated with his head above the water, the two holes on top of his head opening and closing quickly as he breathed in and out. His pectoral flippers were gently flapping. He looked at me and made some chirping and clicking sounds.

"What are you trying to tell me, little whale?"

He tried to speak with me again. This time he grunted, then he repeated the grunts and turned away from me.

For a second my heart dropped to my feet. I thought he was going to swim away again.

"Grayson, don't go. We'll find your mother. Be patient. Sometimes you just have to believe. Sometimes that belief gets you where you want to go, sometimes it carries you a little closer, and then you discover another way."

Grayson lay on his side. He looked tired. Waves were washing over his massive head; he was looking at me and he was listening to the water.

Tilting my ear into the water, so I could hear what he was listening to, I listened and heard something I'd never heard before.

It sounded like a hundred high-pitched sparrows singing through a hundred tiny megaphones turned up to the highest volume.

"What's making that sound?" I asked.

I swam to within two feet of him and floated beside him. He looked as if he was anticipating something.

From behind us came a squeak. Grayson lifted his big gray head and he stared across the water. My eyes followed his gaze. There were long winding waves with no beginning or end, just miles and miles of water, endless, ceaselessly moving water.

But I suddenly saw something cutting across the water, a dark dorsal fin moving very fast, at about twenty knots. Then I saw another and another. In a moment there were three more fins, then twenty-five, thirty, forty, fifty, sixty, seventy, eighty, ninety, one hundred ten, one hundred twenty, thirty, forty, sixty, seventy, one hundred eighty. There were two hundred and twenty dolphins, a sea of dark gray fins bobbing up and down as they raced across the sea, speeding toward Long Beach.

The fins were in the middle of their backs and varied in color from black to light gray. They were all outlined in black. The dolphins were between seven and a half and eight and a half feet long. Their bodies were beautifully cylindrical and they had a distinctive hourglass pattern on their sides and long, soft gray beaks. They also had dark lines that ran from behind their beaks to their heads and another black stripe that went from their jaws to their flippers. It was as if nature had outlined them to emphasize their beautifully and elegantly streamlined shape. It took me a few minutes to realize that they were common dolphins.

They were stunning swimmers: Graceful, powerful, explosive, they dolphined through the water with complete efficiency. And they were so agile. In a second they could cut sharply left or right with only the slightest shift in their body positions and a series of quick flicks with their flukes and fins. Like cyclists, they drafted off one another. The lead dolphin worked the hardest, having to push against the resistance of the water; the next dolphins drafted off the ones in front of them and got a chance to rest and recover before they took the lead.

Loud and utterly animated, they whistled and squeaked directions. They swam beak to fluke, minimizing the drag, and when one dolphin cranked up the speed, the others followed. They were chattering in excited high pitches, and it sounded like they were laughing with glee. Their laughs came all the way up from their tails.

I took a breath and put my head under the water.

It sounded like a canary and it began the same way a

male canary starts to sing: with a single tentative note, then a repeat of the note, then a movement up and down the scale and then all at once a burst into song, an aria filled with loud trills. And a second male joined in a duet duel, trying to out sing the first. Multiple voices joined in, adding whistles, squeaks, chatter, grunts, and clicks. It was a dolphin symphony. And it was in surround-sound, coming from all levels of the water column.

The singing came to an end and I thought the performance was over, but it was just beginning.

Two dolphins leaped two or three feet out of the water. They arced through the air. Suspended in the sky, they held their arc, and then they pointed their beaks and punched a hole in the water, entering it so cleanly they would have made an Olympic diver jealous.

Five more dolphins followed, leaping in complete synchrony. Below the water, the song changed to chatter. More voices, excited, joined in. It looked like the dolphins were passing messages to one another like a sports or business team setting up a play.

There was a long pause of nearly half a minute

while everyone moved into position. A loud trill sounded and a group of twelve or thirteen dolphins leaped out of the water simultaneously; another group followed, larger than the first, then a third, overlapping the previous one. More dolphin groups joined in. It was like watching a series of exploding blue, white, and silver fireworks in the late morning sky.

This, I thought, was the grand finale, but I was wrong. The dolphins were just doing their warm-ups, stretching out their fins and flukes.

The dolphins became more creative, fun, expressive, and daring. They sprinted across the water, leaping higher, arcing across the air, diving and doing faster head-first half turns, pirouettes, and wild, out-of-control spins. They wiped out. They slammed down on their sides and backs, and their dolphin friends, watching from the sidelines, were laughing hysterically. I could hear them. Then they started clowning around.

Two dolphins took off, racing against each other; then one leaped out of the water and tail-walked. He scooted across the water's surface standing upright, and a split second later, the other dolphin followed.

He couldn't sustain the tail-walk as long as the first so he landed right on top of the other with a big splash, as if he was doing an intentional cannonball on top of his buddy.

The first dolphin let out a series of clicks and cawing sounds. He nudged the other dolphin. The second dolphin pushed back. They bumped each other, they bumped again harder, they squeaked, chattered, and then sprinted against each other. Their dorsal fins cut through the water like razor blades slicing through canvas. They leaped up side by side and the first dolphin won again. Underwater, he sounded like he was cackling.

Seven more dolphins joined in, but after doing a few tail-walks, they changed the game. They started doing somersaults. They raced across the water, gaining speed, while the dolphins waiting their turns chattered and squeaked like cheerleaders. The dolphins launched into the air, tucked their heads, threw their tails over their heads, and splashed down on their backs or almost completed a full aerial somersault before they hit the water. And when they submerged they squeaked with delight.

Then I felt something moving below me. It was a dolphin, swimming only three feet down. She rolled over onto her back and looked up at me, and then she rolled over again. She turned sharply and circled back with four more dolphins swimming beside her. They were clicking and squeaking and chattering loudly.

They swam right under me and I wanted to reach out and touch them. More dolphins were joining in. They were swimming stacked three on top of one another with a foot of water between them. And they were chattering, as if giving directions to one another. Two dolphins rolled over quickly, flipper over flipper from one side to the other; the hourglass patterns on their bodies spiraled below me.

Time was spinning. I was losing track of time.

Where was Grayson?

Quickly I turned my head. Grayson was only ten feet from me and he seemed to be watching too.

We watched the dolphins playing for nearly five minutes. Then they disappeared and Grayson moved closer.

A flock of brown pelicans patroled the water. They were about fifty yards from us, gliding on their six-foot-long wings, like mini-gliders, cruising six inches above the violet-blue ocean in single file until the lead pelican began flapping his wings fast. The tips of his wings looked like fingers, grabbing hold of the air while the wing blades themselves were pushing hard against the air to gain altitude.

The pelican climbed sharply, moving like a rock climber up a rock face, nearly straight up. Eight pelicans followed. When they reached twenty feet above the ocean, the lead pelican tucked his neck against his chest. His heavy beak and large pouch pulled him forward and he dove fast.

At the very last moment he opened his beak and tucked his wings back. He hit the water with a huge splash; his yellow feathered crown disappeared beneath the water, then he emerged with about half a dozen anchovy in his pouch. He squeezed out the water, tilted his head back, and swallowed the fish whole. His companions followed and caught pouchfuls of anchovy. Seagulls appeared out of nowhere. Squawking loudly and dive-bombing the pelicans, the seagulls attempted to steal fish from them. They tried to intimidate the pelicans so they'd drop the fish, and they tried to steal the fish right out of the pelicans' pouches.

The birds signaled it was breakfast time. From where we were floating Grayson and I could see a massive school of anchovy. Their small bodies were flickering bright silver and gold in the yellow sunlight and they were moving across the water in schools that stretched half a mile into the distance.

The dolphins reappeared, and split into three groups: Some went to the right, others to the left, and some came up from behind. They encircled the fish, working together, and herded the anchovy into a tight ball. Some had breakfast, some continued to play, and I watched them with delight and fascination. They were so bright, social, animated; most of all, it seemed like dolphins just wanted to have fun. They made me smile and remember my first dolphin encounter.

Once when I was training with a friend off Surfside, one beach south of Seal Beach, two common dolphins swam up from behind him.

He had no idea that they were beside him. I had just ridden a wave into shore and I was heading back out when I saw him on the crest of the wave, riding in the froth while inside the transparent green wave were two dolphins, one on either side of my friend. In the green wave their crescent forms were a darker grayish green, and as the wave lifted them it revealed their size. They were huge. They must have been eight feet long and weighed about two hundred and fifty pounds. Each dolphin was within a hand's reach of my friend. They knew exactly what they were doing. I think I saw them smiling through the wave.

My friend must have sensed their presence. He glanced over his left shoulder to see what was beside him. When he saw the huge gray form, he looked startled, and when he looked over his right shoulder, he panicked. His arms were whipping around his head. He was trying to swim through the air, trying to grab hold of anything to get out of the wave and onto shore. I was laughing so hard I started crying.

The two common dolphins kicked out before the wave broke, before they could be thrown onto the beach, but my friend was riding in the white water, spinning his arms in the froth as fast as he could. He pulled off his goggles, turned, and looked back, and then he saw the dolphins: three more had joined them. They were riding in the next wave. He shook his head, and we laughed very hard and immediately went out to play with them.

Grayson had been watching the dolphins closely, and he had been listening to them talking. He had been making his own sounds. A group of five or six dolphins swam to within ten feet of us. They looked right at us. We held one another in our eyes. I laughed. It was so amazing to see them so close, to see the thrust of tail fins, the power generated by that motion, and to feel the pressure of the water against my legs. Grayson felt it too. He stayed beside me.

Curious, the dolphins swam a few feet closer. They squeaked and chattered more loudly, and in a higher pitch, with more excitement in their voices. Grayson clicked slowly and he made a low grunt and some rasping sounds. They sounded different from one another, as if they came from very different countries, but it seemed as if the dolphins were listening under- water, and squeaking and clicking back to Grayson. Maybe the sounds were a foreign language to him, or maybe he understood the feelings or maybe he felt their sonar. In some way they seemed to be communicating and Grayson seemed to be listening and making sounds back to them. Then the dolphins turned and rejoined their herd moving north.

Grayson then swam toward them. His movement through the water looked like the wave his body formed as he swam through the water. He was so strong and fast. He was exquisite, more beautiful than any of the dolphins. He was gentle and friendly, trusting and sweet. And he had become very special to me. Somehow I think he knew that.

He swam back to me. And I couldn't help but think how amazing it was that this baby whale had come to me to ask for help. That he had trusted me, too.

He turned and headed toward the oil rig; I followed, unsure why we were returning to the area. Something had happened to him. Had he gotten a second wind? Had he realized something that helped him come back again with a fresh mind-set?

His tail movements were slow and efficient. I smiled and wished I had a tail like his so I could swim like him.

A smaller herd of Pacific white-sided dolphins passed within twenty-five yards of us. A couple of them swam to within five feet to investigate, then they rejoined their group. There were about twenty in their herd. They were slightly smaller than the common dolphins, with shorter beaks, which were dark, as were their dorsal fins, flippers, and flukes.

The herd turned and swam a little closer, and I noticed that in between the sets of large dorsal fins were smaller ones. There were baby dolphins. They must have been only a few months old, still dependent upon their mothers for survival. The babies were swimming in the adults' slipstreams, getting a free ride; positioned between the adults, they were pro-tected from predators on all sides.

Two scouts swam right under us. They turned over on their sides and I could see the white stripes along their bodies. They looked me directly in the eyes. And I felt like they were looking deep into me. And I think they felt me look back. They squeaked. And I heard more voices. I looked at Grayson. He seemed to be watching and listening. I think I heard him whistle. I hadn't heard him whistle before. Was he trying to communicate with them? Maybe he didn't know he couldn't and maybe because of that he was able to.

The Pacific white-sided dolphins swam off. Had Grayson heard his mother or thought he heard her? Was that why he swam out to the oil rig—to find her? Or had he been lured there by the sounds emanating from the rig, or even by the dolphins? Maybe he had heard the sunfish speak. Maybe he discussed something with the dolphins. Grayson had managed to get me to understand him; had he done the same with the dolphins?

The swim back to the pier was going to be hard. There was no way around it. The wind was gusting to fifteen knots and the sea was breaking into whitecaps. It was hard to find places between the waves to breathe. And I could hear Grayson swimming nearby, his breaths shorter and more frequent. He was tired and hungry, and maybe cold.

I stopped for a moment to refocus. The tide was against us. The current was flowing at about three-quarters of a knot. This wasn't fun. My speed was normally two knots—two and a half miles per hour.

Grayson's speed was at least double that. I wondered if he was getting cold like I did when I waited for slower swimmers. I hoped he was okay. I looked at him. He was about fifty yards ahead of me, his fluke leaving a momentary footprint in the dark blue-gray water.

The swells were growing from one to two feet, and as I swam I felt like I was bouncing on a trampoline on my stomach. Spray off the waves was splashing into my mouth and I was choking on water.

Grayson swam right up to me, within an inch, and he let me touch him. His skin felt rubbery, like a mushroom, and not at all slimy. It gave a little when I touched it. I reached on top and felt his dimples and then I slid my hand under him and smiled. I held the baby whale in my hand. And I felt the life within him, much the same as I had when I held the tiny grunion, but Grayson's life force was so much bigger.

He trusted me enough to let me touch him. We were from two different worlds—two different beings, with two different lives, and yet somehow we understood each other.

"Everything will be okay, Grayson, don't worry, we will figure this out," I promised.

We swam side by side toward shore. And I felt a new energy. Grayson was swimming easier too. And I was catching his slipstream, riding the tiny waves sliding off his long deep gray body.

In the distance, I saw the Long Beach Lifeguard boat traveling toward us at full speed.

nine

The lifeguards motored alongside us. The older life-guard with the dark brown hair came up from the cabin onto the deck and said, "Glad you decided to head back to shore. We've been keeping an eye on you, but with the change in weather and all the boat traffic, it's getting dangerous to be swimming out here without a boat."

He asked me if I'd seen any sign of Grayson's mother. He had some good news. A crew on a commercial fishing boat about twenty miles north of us had spotted a pod of five gray whales swimming off the rocky Palos Verdes Peninsula. They didn't think the pod included our mother whale, but it was a sign that there were other whales in the area. And that gave us hope, and enabled us to inspire each other.

Steve was doing exactly that. I could hear him speaking on the radio on the lifeguard boat. He was excitedly talking with a fisherman who had been casting his line into the water about a half mile south of us, inside the entrance to Huntington Harbor. The fisherman thought he had seen a whale spouting under the bridge near the mudflats.

He couldn't be sure. It could have been a pelican diving into the water to grab a fish. Whales didn't usually swim inside Huntington Harbor.

But there was a chance it was Grayson's mother and so we decided to wait near the end of the pier by the bait shop, hoping that Grayson would stay with us.

The wind was increasing from the southwest and the gray-blue ocean was erupting into a mass of rolling one-foot waves. The lifeguards moved their boat beside us to buffer the bounce of the chop.

By the time we reached the pier, a group of fishermen and parents with their kids were leaning on the pier railings, looking south, scanning the water for a spout or any movement, but it was hard to see any-thing with the waves and glare off the water.

Two fishing boats joined us, and Carl drove over in his small motorboat. They scanned the water with pro-fessional eyes, intently studying the ocean for a sign.

In the background, now and then, people were speaking to one another on the ships' radios.

Steve's voice came through clearer than the others. He said, "A fisherman on the southern jetty thinks he saw something big swimming around the harbor entrance. He thinks it might be her moving in our direction."

In a minute all the people standing on the pier moved to the left side. Some bent way over the rail-ings to see farther into the distance, while others slowly scanned the water, looking for anything moving our way.

Grayson was restless. He was swimming back and forth like a person pacing. He was breathing faster and shorter. Was he trying to be heard through his breaths? The sound traveled at least half a mile into the air. Maybe he was pacing because he was cold and he was swimming back and forth to stay warm. He had far more body fat than I did and his was far denser than mine. I felt cold deep in my muscles. I was shivering. But I was afraid to get out of the water. If I did, it might affect Grayson badly.

From all the experience I had in open-water swimming I knew that it was an incredible lift to swim with someone else. Just having someone beside me made me feel better. At times when I was lagging, having someone there gave me the confidence to continue; it really made all the difference in the world. I didn't want to climb out of the water because I was afraid that Grayson would think I had abandoned him. He might leave before we ever found his mother.

Sensing his unease, I suggested to the lifeguards that we swim to the southern jetty to see if Grayson's mother was there.

The lifeguards thought it would be better if we stayed put. They thought that if the mother whale was near the southern jetty she was probably retracing her footprints. She would most likely return to the place where she thought she had lost him.

I floated on my back and kicked my feet to generate heat. I couldn't get warm. I tried to think of what else we could do. I rolled on my stomach and watched Grayson swim through the rumpled, silvery green water. He was swimming slower than before. He seemed to be more agitated. His movements were more erratic.

I've got to do something, I told myself, but I didn't know what. Just waiting there and watching him wasn't accomplishing anything, but swimming around in circles wasn't accomplishing much either. Maybe if I think very hard his mother will hear me. Maybe she won't know my words but will sense my brain waves. Maybe she will hear my feelings with her sonar. Maybe she will hear me calling her through the water. Sound waves travel faster through the water than they do through air. Maybe brain waves can travel faster and longer through the water. Please, come this way, over here! I shouted with my mind.

Grayson was breathing faster. He was pacing back and forth, as if he expected something to happen.

How much longer would he be patient? How much longer would he stay with us?

"Please hear me, Grayson's mother, somewhere out there. If it's you swimming near the Huntington

Beach jetty, please swim this way: Grayson is here. Your son is here."

I took a breath and put my face into the water. What would we do if we couldn't find her? We couldn't abandon him. But I couldn't bring him home. Who could take care of him? He had to have his mother's milk. What else could he eat?

He was swimming so slowly toward shore again.

Do whales get hypothermia? Do they cool down? Could they die from the cold? Could he shiver and generate heat? Maybe he was sick and growing sicker. Maybe he had been left behind because he couldn't keep up.

"Please swim this way. Please swim toward Seal Beach. Please swim to the pier."

I thought as hard as I could. I didn't know if it would work. I didn't know if anyone could ever know. But I had to try something. You don't have to hear the words to know someone cares about you. You don't need to hear the words to know someone believes in you. You don't need to hear the words to know someone loves you. You feel it; you know it.

Maybe there was a way she would hear me if I just thought more strongly.

I think Grayson heard me. I think he heard my emotions and felt them too. He floated on the surface near me as if waiting for me to signal what we were supposed to do next.

I projected my thoughts: Be patient. Wait. Nothing is all good or all bad. As a problem develops, so does the solution. Rest here. I will tread water beside you. You will be okay. I know it. I feel it. It will all work out.

Tilting my head back and looking up, I noticed that more people were standing on the pier. See all of them up there, Grayson. They're here for you.

It was as if Grayson understood. He looked up. He saw them and he grunted softly.

The people on the pier pressed against the railings, leaning toward the sea.

They were willing his mother to appear. I hoped she could feel the good vibrations coming from all of the people on the pier. Something was drawing them out there; something made them want to help.

My heart beat faster. I felt something change.

And then I heard a mother's voice from the pier, telling her sons that everyone was out there looking for the mother whale. She warned her youngest son, who was about five years old with blond hair and wearing a dark red sweatshirt, not to stand too close to the edge. His older brother, wearing a bright blue sweatshirt, was standing on the other side of his mother.

The little boy in red stepped in front of his mother. He was so close to the edge that I thought he was going to slip under the railing and fall, but his brother caught his hand, and without even noticing, the younger brother said in a sad high-pitched voice, "Did the baby whale lose his mommy? Where did she go?"

"I don't know," his brother said.

"Why did she leave him?"

The older brother said, "I don't know where she went, but let's look for her. Maybe we can find her."

"Okay," said the younger one, slipping his hand into his mother's and seriously staring across the ocean along with his big brother.

And it happened.

We hoped, believed, tried, worked, learned, and

tried again, and then suddenly it happened in a single moment, all that we hoped for and even a little more.

The sea's surface was changing. An underwater current was colliding with the chop and the waves were growing larger, but only in a wide straight line.

"Look over there! I think I see something!" the little boy shouted excitedly.

It had to be. It just had to be.

"I think I see her! I think I see his mommy!" a strawberry blond little girl shouted in a high joyful voice.

People were leaning so far over, trying to see what the little girl saw, that I hoped the wooden railings would hold the weight.

Then someone was shouting, "I think I see her too. Over there!"

People were craning their necks, shielding their eyes with their hands.

Someone else shouted, "Yes. There she is!"

"There she blows!" A fountain of white spray shot out of the water ten feet into the air.

People were laughing, shouting, pointing, clapping, cheering, and squeezing against the south-facing side of the pier. Parents were lifting kids on their shoulders, and older kids were ducking under and weaving in between the adults to get a better view.

There she was, one of earth's most amazing creatures. Swimming toward us.

Grayson took a few quick breaths and dove, and I stuck my head underwater.

There were sounds coming from the distance, sounds I'd never heard before. They were large, intense, so big I could feel them rumbling through the water.

Then there was nothing. No sound. No feeling. Nothing. Just the rushing sounds of my bubbles rolling out of my mouth, past my ears.

I looked for Grayson. He was gone. Had he found her? Had he swum away with her?

Then I heard his mother: She was talking and she had a beautiful voice—a voice that made me laugh and smile.

She was singing, her clicking and chirping strung together. She paused and made a series of sounds, high sounds and low ones and probably so many more at frequencies that were too low for any of us to hear.

There was a pause. And then I heard a second voice. It had to be Grayson. It was. It was Grayson. He had found her! He was clicking and grunting.

What was he saying to her? What was she saying to him? Was he explaining that he had been looking for her for most of the morning? That he was scared, but that some humans had stayed with him and helped him find her?

They had found each other. That was all that really mattered.

Surfacing, I looked up at Steve. He was beaming. For the first time since I met him, Steve was so emotional he couldn't speak. He smiled and shook his head and pressed his index finger into the corner of his eye to brush away a tear.

Grayson and his mother surfaced near the lifeguard boat. Everyone on the pier and in the boats was smiling, laughing, pointing, exclaiming about the beauty of the whales and the magic of seeing the mother and son swimming together.

Grayson and his mother dove and surfaced ten feet from me. I made sure not to move between mother and son, but they swam over to me.

Grayson's mother was enormous, at least forty-five feet long—longer, I think, than the lifeguard boat. She swam slowly past me. I felt tiny beside her. I held my breath and felt powerful energy emanating from her body. Was she speaking to me? Was she using low frequencies, sounds that were too low for me to hear but that I could feel? I treaded water and looked closer.

She had patches of white barnacles on her sides, dimples on her upper jaw, and more barnacles along her chin. There were three long grooves along her throat that allowed her throat to expand when she fed, and I caught a glimpse of her pink tongue. It was longer than my arm and probably weighed more than a ton. She had baleen plates in her mouth. She used these to filter food—amphipods, mollusks, squid, and other little marine animals—out of the water once she reached the Arctic waters and started feeding again.

She turned and swam to within five feet of me. She was massive and it was amazing; she could move so slowly and she was able to gauge her speed and her size. She knew how close she could get without swimming down something as small as me.

She circled back and swam even closer. I was

thrilled to see this magnificent being beside me. She was so big the wave coming off her body pushed me back, but I was compelled to pull closer to her.

She dove deep under me, and I felt the water quickening. I realized she had been swimming under me when we were at the jetty earlier that morning. That's where she'd lost Grayson. She did what any mother would do; she doubled back and retraced the route she had taken with Grayson that morning. She must have panicked, trying to find her baby in the ocean. She took a massive breath of air and spouted. Her *poof* echoed through the pier and her fountain of water was caught by the wind. It showered the people in the boats and they laughed with delight, in awe at her size and sweet nature.

She slipped through the water as if the ocean were part of her being. As if they were one and the same. And as she swam, she made me think of Grayson, how he was a beautiful swimmer too. And how he must have learned from her. In that moment I realized how amazing life is, how filled with unexpected wonders, and how fortunate I was to be in the ocean that day.

With one lift and push of her tremendous fluke the

mother slid through the water. Her footprints were enormous, maybe seven or eight feet wide. I watched Grayson follow her. His footprints were perhaps two feet wide.

She suddenly swam right under me. I took a breath and looked down. All I could see was the gray top of her head. She was only three feet below. I reached and could almost touch her. All in the same moment, I was fascinated, thrilled, and scared beyond belief. I had never swum with anything as big as her in my life. My heart was pounding in my chest.

With two or three thrusts of her fluke she was swimming fast, moving at four or five knots, but she was so big it took three seconds for her head, back, and fluke to pass under my body. She turned abruptly and swam very slowly two feet from me. She was right beside me. For a moment, I touched her cheek. It felt rubbery and rough where there were barnacles. She tilted her head and she was looking into my eyes. There was a glimmer of light in her big brown eye. I felt a connection between us, just as I had with Grayson. She looked at me. I looked at her. We held each other's gaze.

It seemed like she was saying thank you—at least

that's what I felt. I was so elated, hoping, barely able to believe that she was really there.

She swam one more time around us in a circle with Grayson nestled against her side. She seemed to be showing us that she had Grayson now, and everything was going to be all right.

She gently nudged Grayson and he swam closer beside her, up near her head. He made a soft grunting sound. She replied. He said something else; now, looking back, I think it was goodbye.

As strongly as I could think, as strongly as I could feel, I thought and felt, Farewell, Grayson; farewell, Grayson's mom. In a very short time you have shown me things I would never have discovered on my own. You have taught me how to listen and feel and understand without using words. Even if words could reach to eternity there would not be enough to express the way I feel about you.

They swam under the boats and under me, and I just hoped they could feel what I felt for them: You're going far away, but you will always be in my heart and in my dreams. When I think of you I will smile and always remember this day.

ten

Grayson and his mother spouted, and the sun caught their heart-shaped spray just right. There were two rainbows in the spray, side by side, one big and one little. I could see how happy they were to be together again, how excited Grayson was, and how much his mother cared for him.

They swam beyond the lifeguard boat north toward the jetty. Grayson rode in his mother's wide and strong slipstream, and the lifeguards followed in their boat.

Grayson was gone. It was all that I had hoped for. All that I'd spent hours believing could happen. I watched him swimming into the silvery water, cutting effortlessly across the cross currents, growing smaller and smaller by the minute as the sea expanded behind him. I knew I might never see him again. But I knew that there were experiences in a lifetime that no matter where you are, no matter what else happens, you carry them with you forever.

Steve's arms were resting on the pier railing, and his hand was cupped over his eyes to shield them from the blinding glare. He looked down at me and nodded confidently; his mustache curled when he smiled. He knew Grayson would be okay now. He looked happy one moment and sad the next as emotions spiraled around in him.

"What a wonderful morning," I said.

Steve laughed deeply and nodded a few times. He brushed a tear off his cheek, as emotions surged through him.

I felt a new love and respect for him, for his willingness to stop and help the whales, and for his belief that we could find Grayson's mother. He never doubted it and he never gave up.

We watched the whales until they disappeared behind the jetty. As they did, the small group on the pier cheered, and children sitting high on their parents' shoulders waved goodbye to Grayson and his mother.

The lifeguards offered me a ride ashore and I happily accepted. I was freezing cold, tired, and so hungry. Before I jumped back into the water to swim the last few yards into shore, they radioed their friends on the Long Beach Lifeguard boat patrolling near the *Queen Mary*.

They had spotted the whales. Mother and son were swimming right on course, cruising at three or four miles per hour with the waves and sunshine on their backs.

The lifeguards near the *Queen Mary* escorted the whales to the outer edge of Los Angeles Harbor and they made sure the cargo ships sailing in and out of the harbor knew where the whales were swimming.

Grayson and his mother joined a pod of three other whales that were swimming north, bound for Alaska.

My feet were numb when I climbed onto the beach, but the sand was soft and warm between my toes. As I walked up the beach, bent over and shivering, I wiped the brownish green plankton off my face, then crossed the parking lot and asked the lifeguards who were now on duty in the Seal Beach lifeguard station if I could call home quickly to let my parents know that my workout had taken me a few hours longer than I expected and that I was on my way home.

Later, when I sat down at the breakfast table and ate, I told my mother and father what had happened that morning. I told them I had swum with a baby whale and that friends had helped him find his mother. I didn't make a big deal about it.

Many years have passed now.

When I work out along the California coast I often look for gray whales swimming by. I always wonder what became of Grayson. Did he swim across rough Arctic seas? Had he basked in the warm lagoons off Baja? Had he grown into an adult?

By now, he would be more than thirty years old, and he would still be growing and gaining strength. If he is lucky he will live to be fifty years old. Had he found a mate? Did he have babies of his own?

In winter, spring, and fall, when I'm swimming in the ocean, and I see whales migrating up or down the California coast, I imagine Grayson is swimming with them. He's out in front, full of power, strength, and song. He's using his sonar, guiding the other whales, telling them about the places he's been, the distant seas and far-off shores. These are waters where I've never been, oceans where only gray whales can swim. What would it be like to travel with them?

As the gray whales pass me, I watch them move together across the water and I feel the same awe and wonder as the day I met Grayson.

Sometimes for a moment or two I feel something in the water, a sudden stirring, a high energy force, like the morning Grayson swam with me, and I watch the whales swimming effortlessly across the water, beauty in motion, heading for the distant horizon.

EPILOGUE

When I wrote *Grayson* I initially wrote the story for adults. I didn't realize then that young people would read *Grayson*, too, and connect with his amazing journey. Over the past few years I've heard from teachers and professors who have used *Grayson* in their classes, and I've heard from many other people about their close encounters with animals in the ocean.

It is a great gift to be able to share a story, but I think it is a greater gift to hear back from people who have read this book and one way or another been touched or inspired by it. Linda Rasmussen, who teaches at Highland Ranch Elementary School, read *Grayson* with her third-grade class. To understand the size of the baby whale, her students drew an eighteen-foot whale that now wraps around the classroom, and wrote their own poems and stories based on their connection to the book. Jan Battle and Rick Hodge, who teach high school students in a class run by Bucks County Schools Intermediate Unit in Pennsylvania, told me how they read *Grayson* with their students and then gave them free rein to use it as an inspiration for their own studies. Jan and Rick sent me copies of the students' amazing work: One of them, Chuck, did an extensive environmental project on the Exxon Valdez spill; Stacey and Crystal created a poster that described and identified a large variety of whales; Dan did an incredible report on the Mola mola fish. Rob created an aquarium with three windows, each containing a different type of sea creature. Susie designed and drew a beautiful book jacket. Tim rewrote the last chapter of *Grayson* in a very imaginative way, Jarmal wrote an epilogue in Grayson's voice, and Tom created an interactive "whale phone" that gave his friends a chance to try out whale communication. There are no limits on human ability or imagination, and it is amazing how far these students have taken their own explorations of the natural world.

There have been so many moments when I've been amazed by people's reactions to *Grayson*. Sometimes men I've never met will tell me about heartfelt experiences with their mothers; sometimes mothers will tell me how important their children are to them. What this proves to me is that on some very elemental level we are all connected to each other and to the life around us.

Many people also want to share stories about their own encounters with marine animals. A group of four- and five-year-old Blue Birds in Long Beach, California, who had read *Grayson* with their parents, talked excitedly about going out on boats to watch gray whales migrate along the California coast. They even made me a toy-sized stuffed whale they named Grayson, and each girl signed the whale's belly. Cody, my yellow Lab, proudly carries Grayson around in his mouth on our morning walks. Ninety-eight-year-old Ione Rice, from Torrance, California, read *Grayson* and recounted her enchanting experiences of swimming with dolphins off the coast of Florida. Cindy Hunter, an expert on coral reefs at the University of Hawaii, told me about a more intimidating encounter: She had jumped over the side of her boat to investigate a coral reef and come face to face with a nine-foot-long hammerhead shark. The shark looked right at Cindy, shook its head from side to side, and swam off. When Cindy checked with her shark-expert friends, she learned that the shark had shaken its head to show that it wasn't interested in Cindy; if it had flashed its white belly, though, that would have been an aggressive message telling her to watch out.

One of my favorite stories came from Douglas Chatwick, an author and wildlife biologist. After a radio interview he told me that he was once swimming in one of the lagoons in Baja, Mexico, where female gray whales give birth and teach their babies how to swim before making the long journey to Alaska. A mother whale saw him, picked up her baby whale, rolled over onto her back, and placed the baby on her stomach to display to Douglas.

A friend, Suzy Sullivan, told me about her whale-watching trip off the coast of Maui. She was in a glass-bottom boat when a juvenile humpback whale swam over and started playing with the boat. He swam to one side of the boat and all the people ran to see him, and then he went to the other side so that the people would run over there. Suzy climbed below and lay down on the boat's glass bottom, watching the whale swim so close to the glass that he touched it. He swam to her. They were cheek to cheek. She said it was like being kissed by a whale.

Grayson has been translated into eleven different languages. While doing research for the Brazilian Portuguese edition I met John Calambokidis, a research biologist at Cascadia Research who is an expert on gray, blue, and humpback whales. In the course of his scientific work he has gotten an incredible window into the lives of whales:

My work with Cascadia Research involves tracking movements of known individual gray, blue, and humpback whales based on their natural markings (photographic identifications). That has proven to be a very powerful scientific tool for tracking population size, trends, and migrations. On a nonscientific note, this has also allowed a more direct and personal connection to individual whales and familiarity with them as individuals. On many occasions I have seen the same whale off the U.S. West Coast in the summer and fall that I then see near Central America in winter (sometimes just one to two months apart).

We have had a number of encounters with friendly humpback whales, including one in which two whales ended up lifting my small inflatable boat up in the air on their backs and spinning me around with their pec fins.

These encounters have given me a chance to know whales as individuals and work with them in ways that have given me a great appreciation of their uniqueness, size, and gentleness.

While the "whale phone" that Tom from Bucks County created may seem like a far-fetched idea, the truth is that a real whale phone is being designed by Jim Jolly, an electrical engineer in Hawaii. In designing a listening device to detect tsunamis that could put Hawaii at risk, Jim and his colleagues have used the phone cable that AT&T created to link communications between the continental United States and Hawaii. But in the process, Jim and the other designers discovered that the phone cable also serves as a deep-sea party line that lets them listen in on whale communication. If you want to listen, too, you can go to the following Web site: http://www.soest.hawaii.edu/GG/DeepoceanOBS/.

And as Grayson continues to swim out into the world, I'm sure there will be many more tales to share with you about how his story has inspired creativity and prompted others to share their magical experiences with the creatures of the sea.